Historical Developments in Singular Perturbations

Robert E. O'Malley

Historical Developments in Singular Perturbations

 Springer

Robert E. O'Malley
Department of Applied Mathematics
University of Washington
Seattle, WA, USA

ISBN 978-3-319-36382-0   ISBN 978-3-319-11924-3 (eBook)
DOI 10.1007/978-3-319-11924-3
Springer Cham Heidelberg New York Dordrecht London

Mathematics Subject Classification: 34D15, 74G10, 81Q15

© Springer International Publishing Switzerland 2014
Softcover reprint of the hardcover 1st edition 2014

Printed on acid-free paper

Springer is part of Springer Science+Business Media (www.springer.com)

But is knowledge increasing or is detail
accumulating?

Jeanette Winterson,
Sexing the Cherry

Preface

The author wishes to describe the development of singular perturbations, including its history, accumulating literature, and its current status, as he (somewhat personally) understands it after nearly 50 years' study. The word "development" is intended to invoke *development* as both a French equivalent for series and an historical advance in the subject. The presentation is aimed at early graduate students in applied mathematics and related fields, so there is little emphasis on rigor. You are encouraged to alter this point of view, if you wish, by consulting the cited references. Sophisticated approaches are mentioned, but generally not detailed. The author's hope, indeed, is that this presentation will be accessible and of interest to a broad audience. Readers are expected to be curious and will need a working knowledge of elementary ordinary differential equations, including some familiarity with power series techniques, and of some advanced calculus.

The monograph's first concern is fluid dynamical boundary layers, emphasizing the critical role played by their inventor, Ludwig Prandtl, a professor at the University of Göttingen from 1904 until 1947. Prandtl contributed extensively to aerodynamics. The institutes he founded and many he trained supported Germany in World War II. The anti-Semitism and other negative aspects of the Nazi regime, however, forced many to leave Göttingen and Germany. One result was to make singular perturbations a subject of increasing international activity, important in the broader subsequent development of applied mathematics as a separate discipline.

Prandtl's inner and outer approximations generalize to the more formalized method of matched asymptotic expansions, a procedure that has been extensively applied with great success throughout the sciences, as well as to the construction of uniformly valid composite expansions that relate to the ideas of A. N. Tikhonov and N. Levinson, conceived by less-applied mathematicians about 50 years after Prandtl's discovery. The resulting series generally diverge, so the concept of an asymptotic approximation, independently defined by T. Stieltjes and H. Poincaré in 1886, is central. We benefit from generalizing convergent approximations, classically applied to linear equations, by utilizing stretched variables. Unlike the usual situation, we will typically be able to *construct* explicit asymptotic approximations for a variety of challenging problems. Overall, we seek to provide the reader an overview that will determine the asymptotic behavior.

Multiscale phenomena now seem universal. They generally require special computational and analytical methods. Thus, it is not surprising that multiscale asymptotic techniques can provide efficient ways to analyze nonlinear oscillations on long time intervals as well as handle typical boundary and shock layers where solutions feature narrow regions of rapid change. Much of our progress relies on the appropriate ultimate introduction of slowly varying amplitudes. We will illustrate the underlying ideas and the variety of limiting solutions possible by studying simple examples. Many of those are pearls found in the literature cited, consisting of over 500 references. They comprehensively well represent the important study being pursued. Moreover, the large bibliography encourages readers to consider the literature broadly and to study far beyond the limits of this monograph.

Many predecessors deserve credit for teaching us singular perturbations. These include colleagues and students. Indeed, friends worldwide. The author apologizes to scholars whose work he failed to mention. The extent of his omissions is obvious from his lack of extended reference to Joseph B. Keller (1923-), whose abundant lifetime achievements throughout applied mathematics are centered on cleverly applying asymptotic methods. The author thanks others who have helped him prepare the manuscript, including the reviewers and Olga Trichtchenko and Natalie Sheils, but especially Frances Chen and Alex Goodfriend. He also wishes to express his personal love and thanks to his longtime supportive wife, Candy.

Seattle, WA, USA   Robert E. O'Malley
2014

Contents

Chapter 1

Ludwig Prandtl's Boundary Layer Theory

Boundary layer theory formally came into existence in Heidelberg, Germany at 11:30 am on August 12, 1904 when Ludwig Prandtl (1875–1953), a professor (and chair) of mechanics at the Technical University of Hanover (the youngest professor in Prussia according to Bodenschatz and Eckert [49]), gave a ten-minute talk to the Third International Congress of Mathematicians entitled "Über Flüssigkeitsbewegung bei sehr kleiner Reibung" (On Fluid Motion with Small Friction). (Figs. 1.1 and 1.2). (Recall that Hilbert presented his famous list of twenty-three problems for the new century at the second ICM in Paris in 1900.) Prandtl's resulting seven-page paper in the proceedings, Prandtl [399] [translated to English as NACA Memo No. 452 (1928)] states:

> The physical processes in the boundary layer (Grenzschicht) between fluid and solid body can be calculated in a sufficiently satisfactory way if it is assumed that the flow adheres to the walls, so that the total velocity there is zero—or equal to the velocity of the body. If the velocity is very small and the path of the fluid along the wall is not too long, the velocity will have again its usual value very near to the wall. In the thin transition layer (Übergangsschicht) the sharp change of velocity, in spite of the small viscosity coefficient, produces noticeable effects.

(Here, and below, quotations will usually be indented.)

Prandtl's amazing scientific insight, evolving from an ultra-practical era of hydraulics, but less than a year after the Wright brothers' flight, involves the basic concept of what would become singular perturbations. The governing

© Springer International Publishing Switzerland 2014
R. O'Malley, *Historical Developments in Singular Perturbations*,
DOI 10.1007/978-3-319-11924-3_1

VERHANDLUNGEN

DES DRITTEN INTERNATIONALEN

MATHEMATIKER-KONGRESSES

IN HEIDELBERG VOM 8. BIS 13. AUGUST 1904.

HERAUSGEGEBEN VON DEM SCHRIFTFÜHRER DES KONGRESSES

DR. A. KRAZER

PROFESSOR AN DER TECHNISCHEN HOCHSCHULE KARLSRUHE I. B.

MIT EINER ANSICHT VON HEIDELBERG IN HELIOGRAVÜRE.

LEIPZIG,

DRUCK UND VERLAG VON B. G. TEUBNER.

Figure 1.1: Title page: *Proceedings, Third International Congress of Mathematicians*

Über Flüssigkeitsbewegung bei sehr kleiner Reibung.

Von

L. PRANDTL aus Hannover.

(Hierzu eine Figurentafel.)

In der klassischen Hydrodynamik wird vorwiegend die Beweg⟨ der *reibungslosen* Flüssigkeit behandelt.   Von der *reibenden Flüssig* besitzt man die Differentialgleichung der Bewegung, deren Ansatz du physikalische Beobachtungen wohl bestätigt ist.   An Lösungen di⟨ Differentialgleichung hat man außer eindimensionalen Problemen, sie u. a. von Lord Rayleigh*) gegeben wurden, nur solche, bei de⟨ die Trägheit der Flüssigkeit vernachlässigt ist, oder wenigstens k⟨ Rolle spielt.   Das zwei- und dreidimensionale Problem mit Berü sichtigung von Reibung *und* Trägheit harrt noch der Lösung.   ⟨ Grund hierfür liegt wohl in den unangenehmen Eigenschaften Differentialgleichung.   Diese lautet (in Gibbsscher Vektorsymbolik

$$\varrho \left(\frac{\partial v}{\partial t} + v \circ \nabla v \right) + \nabla \left(V + p \right) = k \nabla^2 v$$

(*v* Geschwindigkeit, ϱ Dichte, *V* Kräftefunktion, *p* Druck, *k* Reibur konstante); dazu kommt noch die Kontinuitätsgleichung: für ink⟨ pressible Flüssigkeiten, die hier allein behandelt werden sollen, ⟨ einfach

$$\operatorname{div} v = 0.$$

Der Differentialgleichung ist leicht zu entnehmen, daß bei nügend langsamen und auch langsam veränderten Bewegungen Faktor von ϱ gegenüber den andern Gliedern beliebig klein wird, daß hier mit genügender Annäherung der Einfluß der Trägheit ⟨ nachlässigt werden darf.   Umgekehrt wird bei genügend rascher

---

*) Proceedings Lond. Math. Soc. 11 S. 57 = Papers I S. 474 f.

**) $a \circ b$ skalares Produkt, $a \times b$ Vektorprodukt, ∇ Hamiltonscher D

rentiator $\left(\nabla = i \frac{\partial}{\partial x} + j \frac{\partial}{\partial y} + \mathfrak{k} \frac{\partial}{\partial z} \right)$.

Figure 1.2: First page: L. Prandtl's talk to International Congress, 1904

Navier–Stokes equations

$$\rho_0(\partial_t u + (u \cdot \nabla)u) + \nabla p = \frac{1}{\text{Re}}\Delta u, \quad \nabla \cdot u = 0$$

(cf., for background, Anderson [8], and Drazin and Riley [125]) reduce to Euler's equations when the viscosity (1/Re for the dimensionless Reynolds number) is zero, but the solutions do not *uniformly* reduce to those of Euler's equations when the viscosity tends to zero. As Ting [483] summarized:

> Prandtl's boundary layer theory initiated a systematic procedure for joining local and global expansions to form uniformly-valid approximations.

This technique, later known as the method of "matched asymptotic expansions," identifies the (local) boundary layer solution and the inviscid solution with the leading order "inner" and "outer" solutions (cf. Darrigol [110], Meier [314], Eckert [131], and O'Malley [373] for background material).

Prandtl's student (and brother-in-law) Ludwig Föppl gave the following opinion in his personal memoirs:

> In view of the importance of the work, I would like to point out its essentials. By that time, there had been no theoretical explanation for the drag experienced by a body in a flowing liquid or in air. The same applies to the lift on an airplane. Classical mechanics was either based on frictionless flow or, when friction was taken into account, mathematical difficulties were so enormous that hitherto no practical solution had been found. Prandtl's idea that led out of the bottleneck was the assumption that a frictionless flow was everywhere with the exception of the region along solid boundaries. Prandtl showed that friction, however small, had to be taken into account in a thin layer along the solid wall. Since that time, this layer has been known as Prandtl's boundary layer. With these simplifying assumptions, the mathematical difficulties ... could be overcome in a number of practical cases.

(See Stewartson [474], regarding the troublesome d'Alembert's paradox (having the implausible no drag conclusion) that Prandtl's theory eliminated).

Prandtl's father Alexander was a surveying and engineering professor at a Bavarian agricultural college in Weihenstephan, while his mother, the former Magdelene Ostermann, spent much of her life as a mental patient. Ludwig was the only child of three to survive birth. Both parents died before he was twenty-five. Though the family was Catholic, Prandtl didn't practice his religion as an adult. His earlier education was in engineering at the Technical University of Munich, but Prandtl received his Ph.D. in mathematics in 1900 from the University of Munich for a thesis on the torsional instability of beams with an extreme depth-width ratio, under the supervision of Professor August Föppl of the Technical University. Prandtl worked for a year at Maschinenfabrik Augsburg-Nürnberg designing diffusers to increase the efficiency of wood-cutting machines, so the flow concerned consisted of wood shavings.

(We note that informative biographies of many mathematicians, though not Prandtl, are available on the MacTutor History of Mathematics archive [355] (St. Andrews University website.)

Officially named the Georg-August Universität Göttingen, the University of Göttingen was founded in 1734 by George II, King of Great Britain and Elector of Hanover. The geometer Felix Klein (1849–1925) hired both Prandtl and Carl Runge at Göttingen in fall 1904 from Hanover to begin an Institute for Applied Mathematics and Mechanics, as he sought to narrow the gulf between mathematics and technology in Göttingen. (Klein's health had collapsed after intense competition with Henri Poincaré concerning automorphic functions, after which he became a university administrator and power broker in German mathematics (cf. Rowe [425], Hersh and John-Steiner [204], Gray [182], and Verhulst [502]). Klein came to Göttingen from Leipzig in 1886, hired David Hilbert from Königsberg in 1895, and earned the honorific title *Geheimrat* (similar to the British privy councillor). Klein especially appreciated Prandtl's

> strong power of intuition and great originality of thought with the expertise of an engineer and the mastery of the mathematical apparatus.

Anticipating more recent schemes, Klein founded the Göttingen Society for the Promotion of Applied Mathematics and Physics, allowing him to obtain and spend supplementary industrial funding for favored projects. Prandtl ultimately supervised 85 dissertations, published 1600 pages of technical papers, and directed an aeronautical proving ground (from 1907) and the Kaiser Wilhelm Institute for Fluid Mechanics (from 1925). Both organizations survive, though with changed names (cf. Oswatitsch and Wieghardt [383]). Prandtl ultimately published about 33 papers on boundary layers and turbulence and directed 31 theses on those subjects.

Theodore von Kármán (1881–1963) arrived as a graduate student in Göttingen in 1906, after receiving an undergraduate degree in Budapest in 1902. Gorn [177] reports:

> Almost from the start, a thinly concealed rivalry developed between the 31-year-old mentor and the 25-year-old pupil. The Hungarian's joie de vivre contrasted sharply with the shy, formal, pedantic habits of Prandtl.

von Kármán received his doctorate in 1908 for work on the buckling of columns and he served as an assistant to Prandtl (Privat-docent) until 1913, when he succeeded Hans Reissner as Professor of Aeronautics and Mechanics at the Technical University of Aachen. He served in the Austro-Hungarian army from 1915–1918 and left Aachen for the California Institute of Technology (or, more informally, *Caltech*) in 1930. In the (posthumous) autobiography von Kármán and Edson [241], von Kármán observes:

> I came to realize that ever since I had come to Aachen, my old professor and I were in a kind of world competition. The competition was gentlemanly, of course. But it was first-class rivalry nonetheless, a kind of Olympic games, between Prandtl and me, and beyond that, between Göttingen and Aachen. The "playing field" was the Congress of Applied Mechanics. Our "ball" was the search for a universal law of turbulence.

The competition between von Kármán and Prandtl regarding turbulence is further highlighted in Leonard and Peters [285].
     von Kármán's former student Bill Sears (cf. Sears [444]) wrote:

> Dr. von Kármán was a master of the *à propos* anecdote. He never forgot a joke, and always had one to illustrate most vividly and tellingly any situation in which he found himself. The result is that memories of von Kármán tend to become collections of anecdotes.

Somewhat unfortunately, then, von Kármán's colorful stories provide us information about Prandtl, though they may not necessarily be strictly accurate. (See Nickelson et al. [348], a recent biography of von Kármán.) von Kármán characterized Prandtl's life as

> particularly full of overtones of naïveté.

One often-quoted von Kármán story (reproduced in Lienhard [289]) reads:

> In 1909, for example, Prandtl decided that he really ought to marry; but he didn't know how to proceed. Finally, he wrote to Mrs. Föppl, asking for the hand of one of her daughters. But which one? Prandtl had not specified. At a family conference, the Föppls made the practical decision that he should marry the eldest daughter, Gertrude. He did and the marriage was apparently a happy one. Daughters were born in 1914 and 1917.

Klaus Gersten, editor of the latest edition of Schlichting's *Boundary Layer Theory* [438], insists that Prandtl married the appropriate daughter since he was 34 and the two Föppl daughters were then 27 and 17.
     Intrigued readers might recall von Neumann's definition (cf. Beckenbach [34]):

> It takes a Hungarian to go into a revolving door behind you and come out first.

See Horvath [216] and Dyson [129], however, regarding the extraordinary (largely Jewish) Hungarian contributions to twentieth-century mathematics.

In his history of aerodynamics, von Kármán [239] summarized Prandtl's skills as follows:

> His control of mathematical methods and tricks was limited....
> But his ability to establish systems of simplified equations
> which expressed the essential physical relations and dropped the
> nonessentials was unique. ...Prandtl had so precise a mind that
> he could not make a statement without qualifying it. This is a
> mistake. To be effective a teacher must see that a beginner in
> science grasps the basic principle before he can be expected to
> understand the exceptions.

In an obituary of Prandtl (Busemann [61]), his former assistant Adolf Busemann, from NASA's Langley Research Center, wrote:

> According to his aim to make his sentences foolproof, they re-
> quired a rewording, a re-phrasing, an explanation of the reasons
> for re-phasing and perhaps a discussion of some exceptions to
> the stated rule. It is quite obvious how different the results of
> such lectures might have been for beginners and for listeners with
> background. Seeing the details as clearly as Prandtl did, and
> never trying to circumvent difficulties by omitting them, made
> the lectures, seminars, colloquia a rich source of information for
> all his pupils, beside those fortunate few who walked with him
> home from work.

Certainly, Prandtl was not a great teacher and expositor (as his father-in-law (cf. Holton [210]). Likewise, there is no doubt that von Kármán was more charming and mathematically sophisticated (cf. von Kármán and Biot [240] and von Kármán's [238] Gibbs lecture to the American Mathematical Society). Sears [444] wrote:

> von Kármán never identified himself as a mathematician...always
> as an engineer. But it was clear to those of us who worked close to
> him that mathematics—applied mathematics—was his first love.

In his autobiography, von Kármán concluded:

> In my opinion Prandtl unravelled the puzzle of some natural phe-
> nomena of tremendous importance and was deserving of a Nobel
> prize.

G. I. Taylor (1886–1975), the leading British fluid dynamicist of the twentieth century, wrote Prandtl in 1935 to say that it was insulting that Prandtl hadn't been awarded a Nobel prize. (Details regarding Prandtl's and Taylor's nominations for a Nobel Prize (in physics) are given in Sreenivasan [471]).

Taylor had, no doubt, promoted the honorary doctorate that Prandtl received from the University of Cambridge in 1936. See Batchelor [31] regarding Taylor and his very productive life. In his 1975 obituary of Taylor, Cavendish Professor Brian Pippard begins

> Sir Geoffry Ingram Taylor...was one of the great scientists of our time and perhaps the last representative of that school of thought that includes Kelvin, Maxwell and Rayleigh, who were physicists, applied mathematicians and engineers—the distinction is irrelevant because their skill knew no such boundaries.

In summarizing fluid dynamics in the first half of the twentieth century, Sydney Goldstein [176] observed (in the first *Annual Review of Fluid Mechanics*):

> In 1928 I asked Prandtl why he kept it so short, and he replied that he had been given ten minutes for his lecture at the Congress and that, being still quite young, he had thought he could publish only what he had time to say. The paper will certainly prove to be one of the most extraordinary papers of this century, and perhaps of many centuries.

More specifically, the prominent French fluid dynamicist Paul Germain [168] wrote:

> Prandtl appears to be the first visionary discoverer of what we may, now, call *fluid dynamics inspired by asymptotics*. ... One must stress that forty lines only are sufficient to Prandtl for delivering the essentials of a number of great discoveries: the boundary layer concept itself, the equations which rule it and how they may be used, their self-similar solutions, the basic law that the boundary layer goes like the square root of the viscosity. It is impossible to announce such major achievements in a shorter way.

There were certainly roots of singular perturbations and the boundary layer concept in much nineteenth-century scientific literature. In his fluid mechanics textbook, Prandtl [401], Prandtl called an 1881 paper by a Danish physicist, L. Lorenz, the

> first paper on boundary layers.

Few would agree today. Indeed, the reference has been dropped in the surviving text, Oertel [357]. Van Dyke [492] notes Laplace's and Rayleigh's work on the meniscus, Stokes' work on the drag on a sphere, Hertz's work on elastic bodies in contact, Maxwell's measurement of viscosity, Helmholtz and Kirchhoff's work on circular-disc capacitors, and Rayleigh and Love's work on thin shells. Additional historical perspective can be found in Bloor [47]. Indeed, Bloor [46] includes another list of precedents.

Hans Reissner's late son, Eric, a long-time applied math professor at MIT, used to insist that the *edge effect* in thin shell theory (cf. Reissner [415]) paralleled fluid dynamical boundary layer analysis, though naturally inter-changing the inner and outer terminology. Von Kármán, likewise, reported that the elder Reissner told him:

> People attribute to Prandtl and to you discoveries which neither
> of you ever made

(cf. Friedrichs [159] for further related comments). We note that Frank [157] describes singular perturbations in elasticity theory, using the perspective of operator theory. Also note Andrianov et al. [9] and the classical reference Gol'denveizer [173]. In Bloor [46], the sociologist argues convincingly that boundary layer theory is a *"social construct."*

One fascinating section of Schlichting and Gersten [438] quotes extensively from Prandtl's lectures from the winter semester of 1931–1932 on intuitive and useful (anschlauliche und nützliche) mathematics, describing the oscil-lations of a very small mass with damping. He considered the *asymptotic solution* of the two-point boundary value problem

$$\epsilon y'' + y' + y = 0, \quad 0 \le x \le 1 \tag{1.1}$$

with $y(0)$ and $y(1)$ prescribed and with ϵ being a small positive parameter, reminiscent of the square root of the reciprocal of the physically relevant Reynolds number. Cole [92] considers the corresponding initial value problem (cf. Chap. 3 below). From here onwards, readers should realize that we will *always* take the symbol ϵ to be such a quantity. To emphasize this, we write

$$0 < \epsilon \ll 1.$$

The small parameters we encounter will typically be *dimensionless*, resulting from scaling a physical model (cf. Lin and Segel [291] and Holmes [208]). Our analyses will typically begin as boundary value problems, skipping both the very important initial modeling stage as well as the mathematical solution's ultimate physical reinterpretation.

The differential equation (1.1) can be solved, following Euler, by setting

$$y = e^{\lambda x}$$

where the constant λ satisfies the characteristic polynomial

$$\epsilon \lambda^2 + \lambda + 1 = 0.$$

The series expansions for its two roots

$$-\nu(\epsilon) \equiv -\frac{(1 - \sqrt{1 - 4\epsilon})}{2\epsilon} = -(1 + \epsilon + \dots)$$

and

$$-\frac{\kappa(\epsilon)}{\epsilon} \equiv -\left(\frac{1+\sqrt{1-4\epsilon}}{2\epsilon}\right) = -\frac{1}{\epsilon}(1-\epsilon-\epsilon^2+\ldots)$$

converge for $\epsilon < 1/4$, following the binomial theorem. Linearity implies that the general solution to the differential equation is a linear combination

$$y(x,\epsilon) = e^{-\nu(\epsilon)(x-1)}c_1(\epsilon) + e^{-\kappa(\epsilon)x/\epsilon}c_2(\epsilon) \tag{1.2}$$

for any constants $c_1(\epsilon)$ and $c_2(\epsilon)$. (We shifted the first exponent for later convenience.) To satisfy the boundary conditions, c_1 and c_2 must satisfy the linear system

$$\begin{aligned}
y(0) &= e^{\nu(\epsilon)}c_1(\epsilon) + c_2(\epsilon) \\
y(1) &= c_1(\epsilon) + e^{-\kappa(\epsilon)/\epsilon}c_2(\epsilon).
\end{aligned}$$

Since the system is nonsingular for ϵ small and positive, Cramer's rule uniquely determines

$$c_1(\epsilon) = \frac{\begin{vmatrix} y(0) & 1 \\ y(1) & e^{-\frac{\kappa(\epsilon)}{\epsilon}} \end{vmatrix}}{\begin{vmatrix} e^{\nu(\epsilon)} & 1 \\ 1 & e^{-\frac{\kappa(\epsilon)}{\epsilon}} \end{vmatrix}} \quad \text{and} \quad c_2(\epsilon) = \frac{\begin{vmatrix} e^{\nu(\epsilon)} & y(0) \\ 1 & y(1) \end{vmatrix}}{\begin{vmatrix} e^{\nu(\epsilon)} & 1 \\ 1 & e^{-\frac{\kappa(\epsilon)}{\epsilon}} \end{vmatrix}}$$

and thereby the displacement $y(x,\epsilon)$. (Estrada and Kanwal [144] provides a distributional approach to solving such problems.) Neglecting multiples of $e^{-\kappa(\epsilon)/\epsilon}$ (which we will later formally declare to be *asymptotically negligible*), we get the asymptotic limit

$$y(x,\epsilon) \sim A(x,\epsilon) + B(x,\epsilon)e^{-x/\epsilon}. \tag{1.3}$$

(The tilde, representing asymptotic equality, will be defined in Chap. 2.) Here the *outer expansion*

$$\begin{aligned}
A(x,\epsilon) &\equiv e^{-\nu(\epsilon)(x-1)}c_1(\epsilon) \\
&\sim e^{1-x}(1 + \epsilon(1-x) + \ldots)y(1)
\end{aligned}$$

satisfies the given differential equation $\epsilon A'' + A' + A = 0$ and the terminal condition $A(1,\epsilon) \sim y(1)$ as a formal power series

$$A(x,\epsilon) = A_0(x) + \epsilon A_1(x) + \ldots \tag{1.4}$$

in ϵ with

$$A_0(x) = e^{1-x}y(1),$$

$$A_1(x) = e^{1-x}(1-x)y(1),$$

etc. Setting like coefficients of ϵ to zero in the differential equation and the boundary condition for A, we'd successively need $A_0' + A_0 = 0$, $A_0(1) = y(1)$;

$A_1' + A_1 + A_0'' = 0$, $A_1(1) = 0$; etc. Thus, the limiting solution $A_0(x)$ for $x > 0$ satisfies a first-order differential equation and the terminal condition. We will later learn that the reason that this boundary condition is selected for the outer expansion A is because the characteristic polynomial has a large negative (i.e., stable) root when $\epsilon \to 0$. The related supplemental term $B(x, \epsilon)e^{-x/\epsilon}$ in (1.3) provides *boundary layer behavior* near $x = 0$, i.e. *nonuniform convergence* of the solution y from $y(0)$ to $A(0, \epsilon)$ in an "ϵ-thick" right-sided neighborhood of $x = 0$. Since $(Be^{-x/\epsilon})' = (B' - \frac{B}{\epsilon})e^{-x/\epsilon}$ and $(Be^{-x/\epsilon})'' = (B'' - \frac{2}{\epsilon}B' + \frac{B}{\epsilon^2})e^{-x/\epsilon}$, this *boundary layer correction* will satisfy the given linear equation (1.1) and the implied initial condition if the factor B satisfies the initial value problem

$$\epsilon B'' - B' + B = 0, \quad B(0, \epsilon) = y(0) - A(0, \epsilon). \tag{1.5}$$

Introducing the formal series expansion

$$B(x, \epsilon) = B_0(x) + \epsilon B_1(x) + \dots \tag{1.6}$$

and equating successive coefficients in (1.5), we naturally require that they satisfy

$$B_0' = B_0, \quad B_0(0) = y(0) - ey(1),$$

$$B_1' = B_1 + B_0'', \quad B_1(0) = -A_1(0) = ey(1),$$

etc., which upon integration agrees with the convergent power series expansion in ϵ of the exact solution (presuming $\epsilon < 1/4$), i.e.

$$B(x, \epsilon) = e^{-(\kappa(\epsilon)-1)x/\epsilon}(y(0) - e^{\nu(\epsilon)}y(1)).$$

Simply compare (1.2) and (1.3) for the unique c_js found, neglecting $e^{-\kappa/\epsilon}$. Our termwise solution for A and B will be generalized to the regular perturbation method in the next chapter. Beware, however, the sum (1.3) is not an exact solution (cf. Howls [219]), though asymptotic (a term we'll carefully define in Chap. 2).

Readers should note the important role played by the rapidly decaying function

$$e^{-x/\epsilon}$$

as $\epsilon \to 0^+$ on $x \geq 0$. (See Fig. 1.3.) It converges *nonuniformly* from the value 1 at $x = 0$ to the trivial limit at each $x > 0$ as $\epsilon \to 0^+$ (quite like the limiting idealized discontinuous unit *Heaviside* step function). Setting $\epsilon = 0$ is meaningless when $x = 0$, but it provides the trivial limit obtained for any fixed $x > 0$. Also note how helpful it would be to immediately assume that the asymptotic solution of (1.1) has the additive form (1.3), a plunge that we will confidently take in our final chapter.

More generally, note that the sum

$$e^{-x/\epsilon} + e^{-\epsilon x}, \quad x \geq 0$$

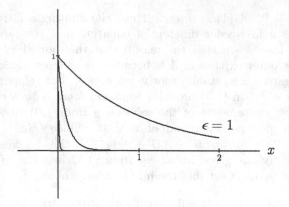

Figure 1.3: The function $e^{-x/\epsilon}$ for $x \geq 0$ and $\epsilon = 1, 1/10, 1/100$

(like its second term) has the power series expansion

$$1 - \epsilon x + \frac{1}{2}\epsilon^2 x^2 + \dots$$

on any interval $0 < \delta \leq x \leq B < \infty$, but the sum converges nonuniformly to 2 at $x = 0$ and to 0 at $x = \infty$.

Finally, observe that Prandtl used the model (1.1) 10 years before K.-O. Friedrichs considered the nearly equivalent problem

$$\nu f_{yy} = \alpha - f_y, \quad 0 \leq y \leq 1, \quad f(0) = 0, \quad f(1) = 1$$

at Brown University's Program of Advanced Instruction and Research in Mechanics (cf. von Mises and Friedrichs [322]). For a fixed constant α, its exact solution

$$f(y, \nu) = (1 - \alpha)\frac{(1 - e^{-y/\nu})}{(1 - e^{-1/\nu})} + \alpha y$$

features an initial layer of nonuniform convergence near $y = 0$ as $\nu \to 0^+$. Indeed, the *outer* limit is $1 + \alpha(-1 + y)$ for $y > 0$ and the nonuniform *initial layer behavior* is described by the correction term $(\alpha - 1)e^{-y/\nu}$ (i.e., by completely neglecting the asymptotically *negligible* term $\epsilon^{-1/\nu}$).

In explaining his basic approach when elected an honorary member of the German Physical Society (upon his retirement), Prandtl [402] stated:

> When the complete mathematical problem looks hopeless, it is recommended to enquire what happens when one essential parameter of the problem reaches the limit zero. It is assumed that the problem is strictly solvable when the parameter is set to zero from the start and that for very small values of the parameter a simplified approximate solution is possible. Then it must be checked whether the limit process and the direct way lead to the

same solution. Let the boundary condition be chosen so that the answer is positive. The old saying *"Natura non facit saltus"* (Nature does not make sudden jumps) decides the physical soundness of the solution: in nature the parameter is arbitrarily small, but it never vanishes. Consequently, the first way (the limiting process) is the physically correct one!

(Quite analogously, Paulsen [387] characterizes a singular perturbation problem by a fundamental difference or change in behavior as a parameter ϵ changes from 0 to a positive number.)

An unusual situation arises when solving algebraic equations that depend analytically on a small parameter. There, singular perturbations occur when the solution involves a Laurent expansion (cf. Avrachenkov et al. [17]).

The extremely slow acceptance of Prandtl's boundary layer theory was noted by Dryden [126] and Darrigol [110]. Bloor [47] explains that the British, in particular, stuck too long to the inviscid fluid dynamics of Rayleigh (from 1876). They were ultimately convinced to accept Prandtl's theory, however. Tani [479] observed:

(i) In the 1905 paper, the essentials of boundary layer theory were compressed into two and a half pages, largely descriptive and extremely curtailed in expression. It is quite certain that the paper was very difficult to understand at the time, making its spread very sluggish.

(ii) Prandtl's idea was so much ahead of the times.

(iii) The genesis of the boundary layer theory stood in sublime isolation: nothing similar had ever been suggested before, and no publications on the subject followed except for a small number of papers due to Prandtl's students for almost two decades.

One might well conclude that boundary layer theory had a slow viscous flow out of Göttingen (cf. O'Malley [373]).

Only a brief mention of boundary layer theory appeared in the fifth edition of the preeminent English-language textbook Lamb's *Hydrodynamics* [278]. There was no mention of the theory in the third and fourth editions of 1906 and 1916. G. I. Taylor [480], however, reported:

When I returned to Cambridge in 1919, I aimed to bridge the gap between Lamb and Prandtl.

Sydney Goldstein's two-volume *Modern Developments in Fluid Mechanics* [175], later reprinted by Dover, was very influential in propagating ideas about the boundary layer. Goldstein had taken on the substantial editorial task of highlighting modern developments upon the death of Sir Horace Lamb in 1934. Lamb finally treated boundary layers extensively in the sixth edition of *Hydrodynamics* [278]. Prandtl's Wilbur Wright Memorial Lecture [400] to the Royal Aircraft Society in 1927 made a substantial impact in Britain, together with the monograph Glauert [170] on airfoils and airscrews (i.e., wing shapes and propellers). Prandtl's student, Hermann Schlichting's 1941–1942 lectures from Braunschweig lived on as mimeographed versions until they were revised and published by G. Braun of Karlsruhe as *Grenzschicht-Theorie* in 1951. Current 8th editions [438] in German and English, at least, are still published by Springer. Schlichting, indeed, became director of Prandtl's proving ground in 1957 (cf. Schlichting [437]).

Before January 11, 1933, when Adolf Hitler was appointed German chancellor by President Hindenburg, Göttingen was a Mecca for mathematicians worldwide (cf. MacLane [304]). Richard Courant (1888–1972) became director of the Mathematics Institute (Wissenschaftlicher Führer) in 1922 (succeeding Klein, who had retired in 1913). The German influence, especially in pure mathematics, had become dominant. Thus, it was natural that the early work of the prominent Japanese mathematician Mitio Nagumo be published in German and that he would visit Göttingen from 1932 to 1934. His lifetime work was greatly influenced by that stay. A chemist's question motivated his 1939 paper on initial value problems for singularly perturbed second-order ordinary differential equations (cf. Yamaguti et al. [530]). As Sibuya observed there,

> when this paper was written, singular perturbations didn't exist.

(The term hadn't been defined.) One 1959 paper by Nagumo (on partial differential equations) also concerns singular perturbations. However, the later critical use of *differential inequalities* to analyze singularly perturbed two-point boundary value problems (cf. Brish [60], Chang and Howes [76], and De Coster and Habets [112]) is directly based on Nagumo's use of upper and lower solutions (or bounding functions) from 1937 onwards. Nickel [347]'s mathematical treatment of boundary layer theory extensively used differential inequalities. The effective use of maximum principles for singularly perturbed differential equations is closely related (cf. Eckhaus and de Jager [138] and Dorr et al. [124]).

Somewhat similarly, the Chinese mathematician Yu-Why Chen came to study at Göttingen in 1928 at the urging of an earlier student of Courant. In a telephone conversation with this author, about 1989, Chen (then living in Amherst, Massachusetts) confirmed that Courant had suggested a thesis topic to him on an ordinary differential equation model featuring boundary layer behavior, because he wished to encourage a mathematical analysis of Prandtl's boundary-layer phenomena. (Chen also said no one had asked him about his thesis in over 50 years.) Chen was granted his doctorate in 1935 for a thesis coinciding with the *Compositio Mathematica* paper, Tschen [485]. His thesis referees at Göttingen were Franz Rellich and Gustav Herglotz, since Courant had been dismissed because he was a Jew. Chen's asymptotics work is largely forgotten, though it overlaps a bit with Wolfgang Wasow's more influential New York University thesis of 1942 (which we will consider in Chap. 3). Such *nostrification*, or not properly recognizing work done elsewhere, may be inadvertent, though it is sometimes claimed to be quite common at certain introverted math centers, however prominent.

Courant had convinced the Rockefeller Foundation's International Education Board (or IEB) to spend $275,000 to build a new Mathematical Institute in Göttingen in 1929, balancing the Institut Henri Poincaré it had already funded in Paris. Just before then, the new Cambridge PhD Sydney Goldstein had visited Prandtl for a year as a Rockefeller Research Fellow. There Goldstein studied numerical boundary layer calculations, a topic extensively developed later by his Manchester and Cambridge colleague, D. R. Hartree.

According to Siegmund-Schultze [460], the IEB's rationale for funding Courant's institute was

> The Board would *not* be interested in ... housing or even helping to house the mathematical department in more agreeable quarters, unless thereby there was a practical certainty that greater and much greater usefulness to a group of sciences would result. In fact, the new mathematical institute was finally erected on Bunsenstrasse in Göttingen close to the physical and chemical institutes and the aeronautical institutes of Ludwig Prandtl. These scientists in time now had better access to the mathematicians and to the mathematical library. Although Prandtl was not involved in the negotiations process, and at one point was mentioned in a slightly disparaging way as "more of an engineer than a physicist" (yet a professor in the University mathematics department), his institutes and the technical sciences are expressly included in the "group of sciences" which were of interest to the IEB's Trowbridge. In fact this was completely in the tradition of Felix Klein who had called Prandtl to Göttingen in 1904 in order to enrich its mathematical environment. Trowbridge's report includes a longer passage on his visit to Prandtl's AVA (aerodynamic proving ground) with its wind tunnel, and mentions the new and more theoretical Kaiser-Wilhelm-Institut.

At the opening ceremony, Hilbert said:

> realizing the idea of building the institute was a great and difficult task entailing various smaller problems. Courant treated each of these with the same love and devotion, always knowing how to find and cajole the most suitable and understanding man for the task

(cf. Bergmann et al. [42]). (Readers may recall that the original, very influential, thousand-page, two-volume *Methoden der mathematischen Physik*

[102] was published in 1924 and 1937 as a collaborative effort at the university, based on Hilbert's lectures, with Courant as editor.)

Applied mathematics was also being simultaneously developed in other German universities, especially Berlin. Quite naturally, Ludwig Prandtl and Richard von Mises established *GAMM*, the German society *Gesellschaft für Angewandte Mathematik und Mechanik*, in 1922 while the sister *Society for Industrial and Applied Mathematics* (SIAM) in the USA wasn't founded until 1952. (The *European Consortium for Mathematics in Industry (ECMI)* didn't start until 1986.)

Erich Rothe was a joint student of Erhard Schmidt and von Mises at Berlin in 1926. Far ahead of the crowd, he subsequently wrote several papers on the asymptotic solution of singularly perturbed partial differential equations and on the skin effect in electrical conductors (cf., e.g., Rothe [422]). Later, while a refugee schoolteacher in Iowa, he published a very good paper on singularly perturbed two-point boundary value problems in the obscure *Iowa State College Journal of Science* (cf. Rothe [423]). (Most of his career was spent at the University of Michigan, where his work included topological degree theory and his first PhD student was Jane Cronin in 1949.)

A major upheaval occurred in Germany on April 7th, 1933. The "Law for the Reorganization of the Civil Service" dismissed Jews from government positions (except those appointed prior to 1914 or who (like Courant) had served in the front lines during World War I (cf. Segal [446] and Siegmund-Schultze [462]). As of 1937, anyone married to a Jew also lost his or her university position. Hitler also targeted others, besides Jews, that he labeled *Untermenschen* (or subhumans). James [225] reports that from 1938 those dismissed were not only forbidden to teach, they were no longer allowed to enter university buildings, including libraries.

After Courant was placed on leave in the spring of 1933, Prandtl joined 27 others in a petition in support of Courant. Petitioners included Artin, Blaschke, Caratheódory, Hasse, Heisenberg, Herglotz, Hilbert, von Laue, Mie, Planck, Prandtl, Schrödinger, Sommerfeld, van der Waerden, and Weyl. Friedrichs later reported:

> Several of the names on the list are those of people who later were considered to be Nazis or near-Nazis, and even at the time some of them were known to be in sympathy with the regime.

Recall that Nazi refers to a supporter or member of Hitler's political party, the National Socialist German Workers' Party or NSDAP. Courant had indeed asked Prandtl to write the Kurator (government representative at the university) on his behalf since he felt Prandtl

> had acted with courage and decision, firing one of his assistants when he discovered that the man was an informer for the Nazi forces

(cf. Reid [411] and Trischler [484]. The assistant was Johann Nikuradse, whose research under Prandtl is still cited.).

From his intermediate stop at Cambridge University in 1934, Courant (according to Siegmund-Schultze [462]) wrote

> Germany's best friends such as Hardy, Flexner, Lord Rutherford, the Rockefeller Foundation get alienated while our institutions, which were unequalled in the world, are destroyed—even Cambridge cannot compare to the old Göttingen. Foreign countries take advantage of the situation and employ people, particularly physicists and chemists, who will in the long run give science and its applications there a big boost.

Ball [24] notes that Courant decided to emigrate when

> My youngest son did not seem able to understand why he should not be in the Hitler Youth, too.

Prandtl protested to the Minister of the Interior

> that the rigid system devised by the race theorists should be flexible enough to allow scholars who were half or a quarter Jewish, who were then logically half or three-quarters German, to be persuaded to join the people's cause.

Mehrtens and Kingsbury [313] report that the number of math students at Göttingen decreased from 432 in 1932 to 37 in 1939. According to Cornwell [99], when Max Planck (as president of the Kaiser Wilhelm Society) expressed his concerns about the deterioration, Hitler replied:

> If the dismissal of Jewish scientists means the annihilation of contemporary German science, then we will do without science for a few years.

(By 1942, the converted Hitler was quoted

> I'm mad on technology.

(cf. Jacobsen [223])). Beyerchen [43] reports that the aging David Hilbert (1862–1943) (who had retired in 1930) when asked at a banquet by the Nazi minister of science

> And how is mathematics in Göttingen now that it has been freed of the Jewish influence?

Hilbert replied:

> Mathematics in Göttingen? There is really none anymore.

Despite his high opinion of Prandtl's scientific merits, G. I. Taylor expressed some personal reservation in his obituary of von Kármán (cf. Taylor [481]):

> By the time the fourth Congress (of Applied Mechanics) was held in Cambridge, England in 1934, the German Jewish members were having a bad time. Theodore was well out of it but was doing a lot for his unfortunate fellow countrymen. Prandtl, who was not Jewish, appeared to be completely taken in by the Nazi propaganda... In 1938, however, when the fifth Congress was organized in Cambridge, Massachusetts by Theodore and Jerome Hunsaker, conditions had changed. Prandtl and my wife and I were staying with Jerome at his home in Boston. The German delegation was strictly watched by political agents who had come as scientist members and Prandtl did not dare to be seen reading American papers. He used to ask my wife to tell him what was in them. After I returned to Cambridge from the fifth Congress I had a letter from Prandtl telling me what a benevolent man Hitler was and including a newspaper cutting showing the Führer patting children's heads. I imagine the poor man did this under pressure from the propaganda machine, for other people told me they had similar letters from him.

Epple et al. [140] reports:

> After the congress, Prandtl tried to convince Taylor in a letter dated October 1938 that Hitler did in fact "turn one million people against himself while eighty million people, however, are his most loyal and enthusiastic followers." The battle Germany had to fight against the Jews, Prandtl continued to explain, was necessary for its self-preservation. He invited Taylor to Germany so that he could see with his own eyes "that we are, in fact, being ruled very well."

Sreenivasan [471] reports that Prandtl and Taylor exchanged 25 letters between 1923 and 1938. Prandtl would usually write in typewritten German (in which Taylor's wife was fluent) while Taylor would reply in handwritten English. In 1933, Prandtl complained about Taylor's handwriting (which is reproduced in Sreenivasan [471])

> Would it be possible for your letters which at the moment are extremely hard to read and cost me a large amount of time be written by someone else who writes more clearly? I hope you don't take umbrage against this remark.

Taylor called Hitler a

criminal lunatic

in his last letter and didn't answer two letters from Prandtl after the Second World War.

Prandtl had hoped that the 1938 or 1942 Congress would be held in Germany, assuming no distinction would be made between Jews and non-Jews. According to Mehrtens and Kingsbury [313], however, the Reich's Ministry of Education wrote Prandtl that

> "Jews of foreign citizenship" who took part would not be regarded as Jews here, but that there would be no place for non-Aryans of German citizenship.

Prandtl accepted the Hermann Göring medal (named for the commander-in-chief of the Luftwaffe or German Air Force and also Hitler's minister without portfolio) in 1939 from the Academy of Aeronautical Research and he chaired an advisory panel for the Air Ministry (Reichsluftfahrtsministerium). Prandtl's institutes grew to have hundreds of workers (cf. Hirschel et al. [207] and Meier [315]). In a letter from 1933, Prandtl wrote

> Hopes increased that, after years of 'unjustified penny-pinching,' the importance of research for the good of the state would fully be recognized.

Trischler [484] reported

> It is then not surprising that the scientists did not regret the passing of democracy, or that they quietly aligned themselves with the new dictatorship, particularly when, with regard to their actual work, virtually no limits were set on their traditional autonomy.

As Epple et al. [140], observed

> we may assume that Prandtl realized, even before the disclosure of the German air force armament plans in the summer of 1935, that the expansion of his institute, at least from the point of view of his official patrons, served to prepare for a new war ... Prandtl thus not only anticipated the actual dynamics of the main phases of war research but also justified some of the research of the prewar arming period not immediately usable by reference to a secondary benefit in the eventuality of a war.

von Kármán reported that after the war, Prandtl said

> he was not a Nazi, but had to defend his country

(cf. von Kármán and Edson [241]). In a manuscript "Reflections of an unpolitical German on the denazification" (1947), Prandtl wrote that he never played a role in politics, but had always served

> State and Science.

His former students, Stanford engineering professors Irmgard Flügge–Lotz and Wilhelm Flügge, to a large extent, defended him (cf. Flügge-Lotz and Flügge [150]). They wrote:

> In the Applied Mechanics Institute, Prandtl was politely squeezed out by a group of flag-waving people, but the Fluid Mechanics Institute started to grow and to enjoy prosperity under the golden rain of government support. A. Betz, while never submitting to party rule, used the interest of the new German airforce to buy expensive equipment and to expand the staff. Among this staff–scientific, technical, and administrative– there was a wide variety of attitudes. Many went with flying colors into the camp of the new masters, and from them the precinct wardens were chosen who had to watch our thoughts and actions and to denounce us if they caught us in a word of doubt or criticism. They were the known enemies, and in their presence people fell silent. For those who did not approve of the regime, there was only the choice between martyrdom and compromise. We do not remember anyone who became a martyr, but the compromise was a walk on a tightrope. No one really knew where the other stood, whether he was a member of the muffled opposition, a spy, or perhaps, at times the one and then the other. This uncertainty, even with regard to former friends, fell like a blight upon the social life.
>
> Prandtl had little understanding for politics and was at times as helpless as a child. He knew that some of the people were like mad dogs, but he could not understand how results of clear logical argument could be rejected furiously if they went against the new doctrine. Standing at the top of the pyramid, he could not avoid giving once in a while a public address, and this was always a nervous strain for the scientific community of the institute. Usually someone had had an advance look at Prandtl's draft of a speech, but who could be sure that he would not be carried away and make some extempore remark that could lead us all into trouble?

The author recalls an earlier seminar by Flügge-Lotz, given in a darkened Stanford classroom with a spotlighted portrait of Prandtl, entitled "The Ludwig Prandtl I Knew."

The biography Vogel-Prandtl [509], by Prandtl's younger daughter Johanna, portrayed him as hostile to the regime, noting that he refused to have a picture of Hitler in his office. Her book also makes public a 1941 letter from Prandtl to Göring which states

> they [the antagonists of "Jewish physics"] have poisoned the air with ... disdain for the past.

This followed efforts by Prandtl to defend Werner Heisenberg and theoretical physics. He summarized by writing:

> In short, it boils down to one thing, namely that a group of physicists, to whom the Führer listens, is raging against theoretical physics and defaming the most meritorious theoretical physicists.

The Nazi physicists and Nobel laureates J. Stark and P. Leonard had attacked "Jewish Aryans" or "white Jews," particularly Heisenberg, the 1933 Nobel prizewinner in physics, because they decided he

thinks like a Jew

(cf. Reeves [410]). The term *Jude in Geiste* (Jewish in spirit) was used. Some mathematicians wrote similar diatribes, especially in the journal *Deutsch Mathematik*, edited by L. Bieberbach and T. Vahlen. Also note Rowe [424].

The Kreisleiter (local Nazi coordinator) in Göttingen wrote in 1937:

> Prof. Prandtl is a typical scientist in an ivory tower. He is only interested in his scientific research which has made him world famous. Politically, he poses no threat whatsoever.... Prandtl may be considered one of those honourable, conscientous scholars of a bygone era, conscious of his integrity and respectability, whom we certainly cannot afford to do without, nor should we wish to, in light of his immensely valuable contributions to the development of the air force

(cf. Ball [24]).

Bodenschatz and Eckert [49] report, based on Prandtl's correspondence,

> By July 1945 the institute was under British administration and had "many British and American visitors." Prandtl was allowed to work on some problems that were not finished during the war and from which also reports were expected. Starting any new work was forbidden.

They also quote a British Intelligence report on a visit to the Kaiser-Wilhelm-Institut (KWI) 26–30 April 1946:

> Much of the equipment of the A.V.A. has been or is in the process of being shipped to the U.K. under M.A.P. direction, but the present proposals for the future of the K.W.I. Göttingen appear to be that it shall be reconstituted as an institute of fundamental research in Germany under allied control, in all branches of physics, not solely in fluid motion as hitherto. Scientific celebrities now at the K.W.I. include Professors Planck, Heisenberg, Hahn and Prandtl among others. In the view of this policy, it is only with difficulty that equipment can be removed from the K.W.I. The K.W.I. records and library have already been reconstituted.

When Courant visited Göttingen in 1947, about the time of Prandtl's retirement, he reported that the aeronautics institute had become a

veritable fortress.

Although ill and depressed, Prandtl was mentally alive. He had given much thought to analog computing machines with a view toward meteorological computations, a longtime interest. As we continue to describe further progress in boundary layer theory, we encourage readers to try to understand Prandtl's behavior in the context of Nazi Germany (cf. Goldhagen [174], Medawar and Pyke [312], and Barrow-Green et al. [30]).

As Flügge-Lotz and Flügge [150] put it:

> The seeds sown by Prandtl have sprouted in many places, and there are now many 'second growth' Göttingens who do not know that they are.

The surviving victims of Hitler's Third Reich (including applied mathematicians) spread worldwide as an intellectual diaspora. Courant's transition from Göttingen to founding what become the Courant Institute at New York University is described in Reid [411]. As in Göttingen, Courant was ambitious to develop as international center at NYU for basic science, ultimately gaining support from the Rockefeller Foundation, the Office of Naval Research, the Atomic Energy Commission, and rich and well-connected individuals. He started anew, with practically no resources. His institute educated many outstanding students and later employed some of them, including the brothers J. B. and H. B. Keller (from Paterson, N. J.), the Hungarian Peter Lax, and the Canadians Cathleen Synge Morawetz and Louis Nirenberg. Among those later involved in asymptotics, Wiktor Eckhaus survived the war in Poland before finishing high school in Holland and a Ph.D. at MIT (cf. Eckhaus [137]); the Hungarian Arthur Erdélyi emigrated from Czechoslovakia to Edinburgh in 1939 with the help of E.T. Whittaker (cf. Colton [94] and Jones [228]), and subsequently split his career between Caltech and Edinburgh; the German/Palestinian Abraham Robinson fled from France to England where he did aeronautical research during the war (cf. Dauben [111]); and Richard von Mises went to Istanbul and then to Harvard (initially, without salary, but with a girl friend!). James [225] reports that of the 2600 Jews assisted by the British Academic Assistance Council, twenty become Nobel laureates, fifty-four were selected Fellows of the Royal Society, thirty-four become Fellows of the British Academy, and ten received knighthoods.

A substantial effort successfully found university positions for most of the émigré mathematicians (cf. Siegmund-Schultze [462]), although there was a counter point of view that the big migration of established academics from Europe to the USA would force some young American mathematicians to become

hewers of wood and drawers of water

(cf. Birkhoff [44] and Joshua 9:23). Einstein (who did not return from America to Germany in 1933) labeled G. D. Birkhoff (1884–1944), the most prominent American-trained mathematician at the time,

> one of the world's great anti-Semites

while Courant said

> I don't think he was any more antisemitic than good society in Cambridge, Massachusetts.

In any case, of the 148 mathematical émigrés after 1933 listed in Siegmund-Schultze [462], 82 came to the United States and all but 7 obtained positions by 1945. As Courant anticipated, many contributed to the allied war effort, including developing the atomic bomb. As Medawar and Pyke [312] conclude:

> The great majority of the scientific emigrants were young and unknown people. Those who later made worthwhile contributions were able to do so because their host countries generally gave them the chance that Germany denied them.

The subsequent prominence of Jewish mathematicians in America is considered in Hersh [203].

K.-O. Friedrichs (1901–1982) left a professorship in Braunschweig in 1937 to come to New York University in order to marry a non-Aryan (cf. Reid [412]). He had been Courant's student at Göttingen, learned fluid mechanics as a postdoc with von Kármán in Aachen, and he encountered boundary layers in plate theory (cf. Friedrichs and Stoker [162]). His lecture notes from the Brown University summer school (cf. von Mises and Friedrichs [322]), his NYU lectures on special topics in fluid mechanics (Friedrichs [160]), and his Gibbs lecture (Friedrichs [161]) all demonstrate his mastery of asymptotics, as part of much broader contributions to analysis and differential equations (cf. Morawetz [326]). In studying nonlinear oscillations, like those described by the van der Pol equation, Friedrichs and Wasow [163] introduced the now universally accepted term *singular perturbations* to distinguish them from the more common situation of a regular perturbation for which uniform convergence implies that a single asymptotic power series suffices. Wasow [516] observed:

> In later years, neither of the two authors could remember to which of them belongs the credit for coining the terminology, but I believe it was Friedrichs.

Wolfgang Wasow (1909–1993) had studied mathematics in Göttingen, seeking to pass his Staatsexamen to become a teacher (cf. Wasow [516] and O'Malley [369]). He passed his orals in 1933 and applied for practice teaching, but was not employed. After some wandering, including teaching at a boarding school in Florence and at Choate School and Goddard College, he took a $600 fellowship at NYU in 1940, arranged by Courant. His 133-page thesis was nearly completed the following summer, under the direction of Friedrichs. It described many singular perturbation examples and made Prandtl's boundary layer theory into a mathematical topic. The work was not immediately well received, however. The first papers he submitted were rejected, but a ten-page paper ultimately appeared in the MIT journal (cf. Wasow [511]),

> with some behind the scenes support from Courant

(according to Wasow [516]). Wasow's autobiography modestly states that his post-thesis research

> was soon overshadowed by two articles by MIT's Norman Levinson who obtained more general results by different methods.

(Levinson's critical work on singular perturbations is summarized in Nohel [353]). Wasow continued to do important work on asymptotics, as summarized in *Asymptotic Expansions for Ordinary Differential Equations* [513] and *Linear Turning Point Theory* [515], as well as much other mathematical and numerical analysis (while working mostly at UCLA and the University of Wisconsin, Madison).

By 1950, singular perturbations were being studied and developed by a variety of mathematicians and engineers worldwide, although hardly in Göttingen. (Curiously, the U.S. Office of Strategic Services recruited over 1600 German scientists (including Wernher von Braun) to come to America after the war (as Operation Paperclip), bleaching any ties to Nazi service (cf. Jacobsen [223]).) Substantial activity was taking place in the Guggenheim Aeronautical Laboratory at Caltech (GALCIT), which von Kármán headed since 1930 (cf. Cole [93]), while many promising efforts underway elsewhere were largely unconnected to the original motivation from fluid dynamical boundary layers.

von Kármán became emeritus at Caltech in 1949, largely due to his expanding duties (since 1944) with the scientific advisory group for the Air Force and (since 1951) with NATO's Advisory Group for Aeronautical Research and Development. In 1962, he received the first National Medal of Science from President Kennedy.

Sears [445] reports:

> He also loved parties, drinks, girls, jokes, the *bon mot*. All his life, he played the part of the dangerous Hungarian bachelor. He succeeded in shocking some of the young wives (and their husbands), but charmed many more. He told us: 'I have decided how I want to die. At the age of 85, I want to be shot by a jealous husband'.

More seriously, he added

> Ulam saw von Kármán and asked John von Neumann who that lit-
> tle old guy was. 'What, you don't know Theodore von Kármán?'
> said von Newmann, 'Why, he invented consulting'.

More generally, the arrival of applied mathematics in academic America is
considered in Richardson [417] and Siegmund Schultze [461]. It spread from
New York, Boston, Providence, and Pasadena, among other centers (cf. Rowe
[426] for a colorful (Truesdellian) description of early applied mathematics at
Brown, including comments on its émigré faculty in the summer of 1942).

Exercises

Although we've not yet accomplished much technically, readers can do the
exercises that follow based on their first course in differential equations (cf.
O'Malley [370]).

1. Find the solution of the initial value problem

$$\epsilon^3 \ddot{y} + \epsilon \dot{y} + y = 1$$

on $t \geq 0$ with $y(0) = 2$ and $\epsilon \dot{y}(0) = 3$ as $\epsilon \to 0^+$.

The answer is an asymptotic solution of the form

$$y(t, \epsilon) \sim 1 + A(t, \epsilon)e^{-t/\epsilon} + \epsilon B(t, \epsilon)e^{-t(1-\epsilon)/\epsilon^2}$$

for power series A and B. (The tilde indicates an asymptotic limit.)

2. Solve the two-point problem

$$\epsilon y'' + y' - (y')^2 = 0, \quad 0 \leq x \leq 1$$

with $y(0) = 1$ and $y(1) = 0$ and describe its boundary layer behavior.

An exact solution is

$$y(x, \epsilon) = -\epsilon \ln \left(1 + e^{-1/\epsilon} - e^{-x/\epsilon} \right).$$

3. (a) Find the exact solution to the problem

$$\epsilon y'' - y' + 2x = 0, \quad 0 \leq x \leq 1$$

with $y(0) = 0$ and $y(1) = 1$.

(b) For small positive values of ϵ, show that

$$y(x, \epsilon) = x^2 + 2\epsilon \left(x + 1 + e^{\frac{x-1}{\epsilon}} \right)$$

up to exponentially negligible terms like $e^{-1/\epsilon}$.

Note that y converges uniformly to x^2, that y' converges nonuniformly near $x = 1$, and that $y''(1)$ is approximately $2/\epsilon$.

4. Show that the two-point problem for the constant coefficient equation

$$\epsilon^2(y'' + ay') - b^2 y = c, \quad 0 \le x \le 1$$

with $b > 0$ has an asymptotic solution of the form

$$y(x, \epsilon) = -\frac{c}{b^2} + A(x, \epsilon)e^{-bx/\epsilon} + B(x, \epsilon)e^{-\frac{b}{\epsilon}(1-x)}$$

where

$$A(x, \epsilon) \sim e^{-\frac{ax}{2}}\left(y(0) + \frac{c}{b^2}\right)$$

and

$$B(x, \epsilon) \sim e^{\frac{a}{2}(1-x)}\left(y(1) + \frac{c}{b^2}\right).$$

Chapter 2

Asymptotic Approximations

(a)   Background

Leonhard Euler (1707–1783), among others in the eighteenth century, was adept at manipulating divergent series, though usually without careful justification (cf. Tucciarone [487], Barbeau and Leah [26], and Varadarajan [493]). Note, however, Hardy's conclusion

> ... it is a mistake to think of Euler as a "loose" mathematician.

As with singular perturbations, the ideas behind asymptotic approximations were not well understood until about 1900. They were presumably unknown to Prandtl in Munich and Hanover. For a classical treatment of *infinite series*, see, e.g., Rainville [404].

See Olver [360] for an example of a *semiconvergent* or *convergently beginning* series. They were defined as follows in P.-S. Laplace's *Analytic Theory of Probabilities* (whose third edition of 1820 is now available online as part of his complete work):

> The series converge fast for the first few terms, but the convergence worsens, and then comes to an end, turning into a divergence. This does not hinder the use of the series if one uses only the first few terms, for which convergence is rather fast. The residual of the series, which is usually neglected, is small in comparison to the preceding terms.

(This translation is from Andrianov and Manevitch [10].)

Throughout most of the nineteenth century, a strong reaction against divergent series, led by the analyst Cauchy, nearly banned their use (especially

© Springer International Publishing Switzerland 2014

R. O'Malley, *Historical Developments in Singular Perturbations*,
DOI 10.1007/978-3-319-11924-3_2

in France). (Augustin-Louis Cauchy (1789–1857) was a professor at École Polytechnique from 1815–1830. Afterwards, he held other positions, sometimes in exile, because his conservative religious and political stances made him refuse to take a loyalty oath.) Note, however, Cauchy [73]. In 1828, the Norwegian Niels Abel (1802–1829) wrote:

> Divergent series are the invention of the devil, and it is shameful to base on them any demonstrations whatsoever. By using them, one may draw any conclusion he pleases and that is why these series have produced so many fallacies and so many paradoxes.

Verhulst [500] has a similar quote from d'Alembert. Kline [255] includes considerable material regarding divergent series. In particular, he points out that Abel continued

> That most of the things are correct in spite of that is extraordinarily surprising. I am trying to find a reason for this; it is an exceedingly interesting question.

Further, Kline quotes the logician Augustus De Morgan of University College London as follows:

> We must admit that many series are such as we cannot at present safely use, except as a means of discovery, the results of which are to be subsequently verified and the most determined rejector of all divergent series doubtless makes use of them in his closet ...

Finally, the practical British engineer Oliver Heaviside wrote

> The series is divergent; therefore, we may be able to do something with it.

In his *Electromagnetic Theory* of 1899, Heaviside also wrote

> It is not easy to get up any enthusiasm after it has been artificially cooled by the rigorists... There will have to be a theory of divergent series.

Attitudes and developments no doubt somewhat reflect the alternations in French and European politics during those turbulent times. *Abel summability* of power series was originated by Euler, but it is usually named after Abel. Roy [427] reports that Abel called the technique

horrible,

saying

Hold your laughter, friends.

In 1886, however, Henri Poincaré (1854–1912) and Thomas Joannes Stieltjes (1856–1894), simultaneously and independently, provided the valuable definition of an asymptotic approximation and illustrated its use and practicality. Their papers were, respectively, published in *Acta Math.* and *Ann. Sci. École Norm. Sup.* (cf. Poincaré [393] and Stieltjes [475]). The latter was Stieltjes' dissertation. Stieltjes called the series *semi-convergent*. Poincaré (a professor at the Sorbonne and École Polytechnique) had studied with Hermite, who was in close contact with Stieltjes, primarily through letters (cf. Baillaud and Bourget [20]). Like most others, we tend to emphasize the special significance of Poincaré because of his important later work on celestial mechanics, basic to the two-timing methods we will later consider (see the centennial biographies, Gray [182] and Verhulst [502]). Stieltjes was Dutch, but successfully spent the last 9 years of his short life in France.

In contrast to convergent series, a couple of terms in an asymptotic series, in practice, often provide a good approximation. This is especially true for singular perturbation problems, as we shall find. McHugh [310] and Schissel [435] connect the topic to the classical ordinary differential equations literature. One's first contact with asymptotic series may be for linear differential equations with *irregular singular points* (cf. Ford [151], Coddington and Levinson [91], or Wasow [513]). An example is provided by the differential equation

$$x^2 y'' + (3x - 1)y' + y = 0$$

which has the formal power series solution

$$\sum_{k=0}^{\infty} k! x^k,$$

convergent only at $x = 0$. The same series arises (as we will find) in expanding the exponential integral, while the series

$$x \sum_{k=0}^{\infty} (-1)^k k! x^k$$

formally satisfies

$$x^2 y' + y = x.$$

The most popular book on asymptotic expansions may be Erdélyi [141]. (My copy cost \$1.35.) The Dover paperback (now an e-book) was based on Caltech lectures from 1954 and was originally issued as a report to the U.S. Office of Naval Research. The material is still valuable, including operations on asymptotic series, asymptotics of integrals, singularities of differential equations, and differential equations with a large parameter. Arthur Erdélyi (1908–1977) came to Caltech in 1947 to edit the five-volume *Bateman Manuscript Project* (based on formulas rumored to be in Harry Bateman's shoebox collection. Bateman was a prolific faculty member at Caltech from 1917 to 1946.) Erdélyi remained in Pasadena until 1964 when he returned to the University of Edinburgh to take the Regius chair that had been held by his hero, E. T. (later Sir Edmund) Whittaker, from 1912 to 1946. (Whittaker wrote *A Course in Modern Analysis* in 1902 and was coauthor with

G. N. Watson of subsequent editions from 1915 (cf. [521]).) Influenced by the work feverishly underway among engineers at GALCIT, Erdélyi and some math graduate students also got involved in studying singular perturbations (one was the author's own thesis advisor, Gordon Latta, whose 1951 thesis [281] had H. F. Bohnenblust as advisor). Erdélyi's *Asymptotic Expansions* was much influenced by E. T. Copson's Admiralty Computing Service report of 1946, commercially published by Cambridge University Press in 1965 [98], and by the work of Rudolf Langer, Thomas Cherry, and others in the 1930s regarding turning points. Much more applied mathematical activity involving asymptotics took place in Pasadena after Caltech started its applied math department about 1962, initially centered around Donald Cohen, Julian Cole, Herbert Keller, Heinz-Otto Kreiss, Paco Lagerstrom, Philip Saffman, and Gerald Whitham, among others contributing to practical applied asymptotics. Their students have meanwhile been most influential in the field.

The historical background to Cauchy's own work on divergent series is well explained by the French mathematician Émile Borel (1871–1956) in Borel ([52], originally from 1928). Borel wrote:

> The essential point which emerges from this hasty review of Cauchy's work on divergent series is that the great geometer never lost sight of this matter and constantly searched this proposition, which he called
>
> a little difficult,
>
> that a divergent series does not have a sum. Cauchy's immediate successors, on the contrary, accepted the proposition with neither extenuation nor restriction. They remembered the theory only as applied in Stirling's formula, but the possibility of practical use of that divergent series seemed to be a totally isolated curiosity of no importance from the point of view of general ideas which one could try to develop on the subject of analysis.

Andrianov and Manevitch [10], among others, report that Borel traveled to Stockholm to confer with Gösta Mittag-Leffler, after realizing that his summation method of 1899 gave the "right" answer for many classical divergent series. Placing his hand on the collected works of his teacher Weierstrass, Mittag-Leffler said, in Latin,

> The master forbids it.

Nonetheless, Borel had won the first prize in the 1898 Paris Academy competition "Investigation of the Leading Role of Divergent Series in Analysis." See Costin [101] for an update on *Borel summability*.

Another important early book [196] on divergent series is by the British mathematician G. H. Hardy (1877–1946), a leading British pure mathematician and a professor successively at both Oxford and Cambridge. It was published posthumously in 1949 with a preface by his colleague J. E. Littlewood saying:

about the present title, now colourless, there hung an aroma of
paradox and audacity.

Hardy's introductory chapter is especially readable, filled with interesting and
significant historical remarks. The book has none of the anti-applied slant nor
personal reticence (cf. Hardy [195], Littlewood [296], or Leavitt [282]) often
linked to Hardy. Overall the monograph is quite technically sophisticated, as
is his related *Orders of Infinity* [194].

Olver's *Asymptotics and Special Functions* [360] includes a rigorous, but
very readable, coverage of asymptotics, with a computational slant toward
error bounds. (British-born and educated, Olver came to the United States
in 1961.) At age 85, Frank Olver (1924–2013) was the mathematics editor of
the 2010 *NIST Digital Library of Mathematical Functions* [361] and of the
associated *Handbook*, the web-based successor to Abramowitz and Stegun [2]
(which originated at the U.S. National Bureau of Standards, the predeces-
sor of the National Institute of Standards and Technology, and which may
have been the most popular math book since Euclid.) It demonstrates that
asymptotics is fundamental to understanding the behavior of special functions, which still remain highly relevant in this computer age.

Among many other mathematics books deserving attention by those wish-
ing to learn asymptotics are Dingle [123], Bleistein and Handelsman [45],
Bender and Orszag [36], Murray [338], van den Berg [40], Wong [525], Ramis
[405], Sternin and Shatalov [473], Jones [229], Costin [101], Beals and Wong
[33], Paris [386], and Paulsen [387]. Readers will appreciate their individual
uniqueness and may develop their own personal favorites.

(b) Asymptotic Expansions

In the following, we will write

(i)
$$f(x) \sim \phi(x) \quad \text{as} \quad x \to \infty \tag{2.1}$$

if $\frac{f(x)}{\phi(x)}$ then tends to unity. (We will say that f is asymptotic to ϕ as
$x > 0$ becomes unbounded.)

(ii)
$$f(x) = o(\phi(x)) \quad \text{as} \quad x \to \infty \tag{2.2}$$

if $\frac{f(x)}{\phi(x)} \to 0$ (Alternatively, one can write $f \ll \phi$.) and

(iii)
$$f(x) = O(\phi(x)) \quad \text{as} \quad x \to \infty \tag{2.3}$$

if $\frac{f(x)}{\phi(x)}$ is then bounded.

We often call these relations *asymptotic equality* and the *little o* and *big O* Landau (or Bachmann–Landau) order symbols (after the number theorists who introduced them in 1894 and 1909, respectively). (Olver [360] calls the O symbol a fig leaf, since the implied bound (which would be very useful when known) isn't provided.) *Warning*: We need to be especially careful when the comparison function ϕ has zeros as $x \to \infty$. The symbol tilde $\sim$ is used to distinguish asymptotic equality from ordinary equality.

As our basic definition, we will use (after Olver): A necessary and sufficient condition that $f(z)$ possess an *asymptotic (power series) expansion*

$$f(z) \sim a_0 + \frac{a_1}{z} + \frac{a_2}{z^2} + \ldots \quad \text{as} \quad z \to \infty \quad \text{in a region } R \qquad (2.4)$$

is that for each nonnegative integer n

$$z^n \left\{ f(z) - \sum_{s=0}^{n-1} \frac{a_s}{z^s} \right\} \to a_n \qquad (2.5)$$

as $z \to \infty$ in R, uniformly with respect to the allowed phase (i.e., argument) of z. The coefficients a_j are uniquely determined (as for convergent series). They're not always the Taylor series coefficients, however. Also note that the limit point ∞ can be replaced by any other point and that (2.5) can be interpreted to be a recurrence relation for the coefficients a_n of (2.4).

An important case, often arising in applications, occurs when the asymptotic expansion with respect to $\frac{1}{z}$ depends on a second parameter, say θ. When the second parameter takes on (or tends to) a critical value θ_c, the expansion may become invalid. The asymptotic expansion is then said to be *nonuniform* with respect to θ.

Convergence factors are sometimes introduced to "make" divergent series converge. Likewise, the Borel–Ritt theorem is often invoked to provide a holomorphic sum to a divergent series (cf. Wasow [513]).

We also note, less centrally, that Martin Kruskal [266] perceptively introduced the term *asymptotology* as the art of handling applied mathematical systems in limiting cases, formulating seven underlying "principles" to be adhered to (cf. the original paper and Ramnath [406]). (They are simplification, recursion, interpolation, wild behavior, annihilation, maximum balance, and mathematical nonsense.)

A very useful elementary technique to obtain asymptotic approximations is the common method of *integration by parts*. We illustrate the technique by considering the *exponential integral*

$$Ei(z) \equiv \int_{-\infty}^{z} \frac{e^t}{t} \, dt, \qquad (2.6)$$

with integration taken along any path in the complex plane, cut on the positive real axis, with $|z|$ large. Repeated integration by parts gives

$$Ei(z) = \frac{e^z}{z} + \int_{-\infty}^{z} \frac{e^t}{t^2} \, dt = \frac{e^z}{z} + \frac{e^z}{z^2} + 2 \int_{-\infty}^{z} \frac{e^t}{t^3} \, dt,$$

etc., so for any integer $n > 0$, we obtain

$$Ei(z) = \frac{e^z}{z}\left(\sum_{k=0}^{n}\frac{k!}{z^k} + e_n(z)\right) \tag{2.7}$$

for the (scaled) remainder

$$e_n(z) \equiv (n+1)!\, ze^{-z}\int_{-\infty}^{z}\frac{e^t}{t^{n+2}}\,dt. \tag{2.8}$$

If we define the region R by the conditions $\text{Re}\,z < 0$ and $|\arg(-z)| < \pi$, so that $|e^{t-z}| \leq 1$ there, we find that

$$|e_n(z)| \leq \frac{(n+1)!}{|z|^{n+1}}, \tag{2.9}$$

i.e. the error after using the first $n+1$ terms in the power series for $ze^{-z}Ei(z)$ is less in magnitude than the first neglected term in the series when $z \to \infty$ in the sector of the left half plane. Thereby, as expected, the series expansion is asymptotic there. (Recall an analogous error bound, and the related *pincer* principle, for real power series whose terms have alternating signs.)

For $n = 1$,

$$Ei(z) = \frac{e^z}{z}\left(1 + \frac{1}{z} + e_1(z)\right)$$

with $|e_1(z)| \leq \frac{2}{|z|^2}$ in the open sector R. For z fixed, the error is bounded. Moreover, we can nicely approximate $Ei(z)$ there by using the first two terms

$$e^z\left(\frac{1}{z} + \frac{1}{z^2}\right)$$

of the sum if we simply let $|z|$ be sufficiently *large*. This is in sharp contrast to using a *convergent* expansion in powers of $\frac{1}{z}$, where we would typically need to let the number n of terms used become large in order to get a good approximation for any *given* z within the domain of convergence.

More surprising is the idea of *optimal truncation* (cf. White [519] and Paulsen [387]). A calculus exercise shows that for any given z, the absolute values of successive terms (i.e., our error bound) in the expansion (2.7) reach a minimum, after which they increase without bound. (Numerical tables for this example are available in a number of the sources cited.) This minimum occurs when $n \sim |z|$, so if this asymptotic series is truncated just before then, the remainder will satisfy

$$|e_n(z)| \leq \sqrt{\frac{2\pi}{|z|}}e^{-|z|} \tag{2.10}$$

when we use *Stirling's approximation*

$$\Gamma(x) = (x-1)! \sim \left(\frac{x}{e}\right)^x \sqrt{\frac{2\pi}{x}} \left(1 + \frac{1}{12x} + \frac{1}{288x^2} + \dots\right) \text{ as } x \to \infty \quad (2.11)$$

for $(|z|+1)!$ (cf., e.g., Olver [360]). The latter series diverges for all x, but gives the remarkably good approximation 5.9989 for the rather small $x = 4$.

Spencer [470] states

> Surely the most beautiful formula in all of mathematics is Stirling's formula ... How do the two most important fundamental constants, e and π, find their way into an asymptotic formula for the product of integers?

Equation (2.11) seems actually to be due to both de Moivre and Stirling (cf. Roy [427]). This error bound for $Ei(z)$ is, indeed, asymptotically smaller in magnitude as $|z| \to \infty$ than any term in the divergent series! Thus, this bound is naturally said to display *asymptotics beyond all orders*.

Paris [386] points out that similar exponential improvements via optimal truncation can often be achieved. He cites the following theorem of Fritz Ursell [489]:

Suppose $f(t)$ is analytic for $|t| < R$ with the Maclaurin expansion

$$f(t) = \sum_{n=0}^{\infty} C_n t^n$$

there and suppose that

$$|f(t)| < K e^{\beta t}$$

for $r \le t < \infty$ and positive constants K and β. Then (using the greatest integer function [ ]), he obtains

$$\int_0^{\infty} e^{-xt} f(t) dt = \sum_{n=0}^{[rx]} \frac{C_n n!}{x^{n+1}} + O(e^{-rx})$$

as $x \to \infty$. Thus, the Maclaurin coefficients of $f(t)$ (about $t = 0$) provide the asymptotic series coefficients for its Laplace transform (about $x = \infty$) (because the kernel e^{-xt} greatly discounts other t values).

We shouldn't extrapolate too far from the example $Ei(z)$ or Ursell's theorem. The often-made suggestion to truncate when the smallest error is attained is not always appropriate. (Convergent series, indeed, attain their smallest error (zero) after an infinite number of terms.) However, we point out that considerable recent progress has resulted using *exponential asymptotics*, by reexpanding the remainder repeatedly and truncating the asymptotic expansions optimally each time (cf. Olde Daalhuis [358] and Boyd [55–57]).

Boyd tries to explain the divergence of the *formal regular power series expansion*

$$u(x, \epsilon) = \sum_{j=0}^{\infty} \epsilon^{2j} \frac{d^{2j} f}{dx^{2j}} \tag{2.12}$$

as an asymptotic solution of

$$\epsilon^2 u'' - u = f(x), \quad -1 \le x \le 1$$

by using the representation

$$u(x, \epsilon) = \int_{-\infty}^{\infty} \frac{F(k)}{1 + \epsilon^2 k^2} e^{ikx} \, dk$$

of the solution as an inverse Fourier transform (cf. Boyd [56]) with F being the transform of f. A critical point is the *finite* radius of convergence of the power series for $\frac{1}{1+\epsilon^2 k^2}$. Boyd seems to be first of many authors to quote Gian-Carlo Rota [421]:

> One remarkable fact of applied mathematics is the ubiquitous appearance of divergent series, hypocritically renamed asymptotic expansions. Isn't it a scandal that we teach convergent series to our sophomores and do not tell them that few, if any, of the series they meet will converge? The challenge of explaining what an asymptotic expansion is ranks among the outstanding taboo problems of mathematics.

In addition to asymptotic power series to approximate a given function, it will often be helpful to use more *general asymptotic expansions*

$$\sum_{n \ge 0} a_n \phi_n(\epsilon).$$

Here, the a_ns are constants and we will suppose that $\{\phi_n\}$ is an *asymptotic sequence* of monotonic functions (or *scale*) satisfying

$$\frac{\phi_{n+1}}{\phi_n} \to 0 \quad \text{as} \quad \epsilon \to 0, \quad n = 0, 1, 2, \ldots, \tag{2.13}$$

generalizing the powers. We will again let the symbol tilde ($\sim$) denote *asymptotic equality*

$$f(\epsilon) \sim \sum_{n=0}^{\infty} a_n \phi_n(\epsilon) \tag{2.14}$$

where for any integer $N > 0$

$$f(\epsilon) = \sum_{n=0}^{N} a_n \phi_n(\epsilon) + O(\phi_{N+1}). \tag{2.15}$$

Often, it will be helpful to limit N and to restrict ϵ to appropriate complex sectors (about, perhaps, the positive real half axis). In the special case of an asymptotic power series, we simply have $\phi_n(\epsilon) = \epsilon^n$. Note that the coefficients in (2.14) are uniquely determined since

$$a_J = \lim_{\epsilon \to 0} \left(\frac{f(\epsilon) - \sum_{n=0}^{J-1} a_n \phi_n(\epsilon)}{\phi_J(\epsilon)} \right) \qquad \text{for each } J. \qquad (2.16)$$

To multiply asymptotic expansions (2.14), it is convenient if the sequence satisfies

$$\phi_n(\epsilon)\phi_m(\epsilon) = \phi_{n+m}(\epsilon)$$

for all pairs n and m. (Determining an appropriate asymptotic sequence $\{\phi_n\}$, to use for a given f arising in, say, some application may not be simple, however. In response, Murdock [335] suggests a method of *undetermined gauges*.) When we let the a_ns depend on ϵ, the series (2.14) is called a *generalized* asymptotic expansion. Their coefficients $a_n(\epsilon)$ are then no longer unique. Such expansions are, nonetheless, commonly used, here and elsewhere.

Some further write

$$f \sim \sum_{n=0}^{\infty} f_n(\epsilon)$$

whenever

$$f(\epsilon) - \sum_{n=0}^{N} f_n(\epsilon) = o(\phi_N(\epsilon))$$

for every N.

An important scale is the *Hardy field* of "logarithmico-exponential" functions, consisting of those functions obtained from ϵ by adding, multiplying, exponentiating, and taking a logarithm a finite number of times.

We note the important fact that a *convergent* series is asymptotic. This follows since the terms $a_k z^k$ of a convergent power series or analytic function

$$f(z) = a_0 + a_1 z + a_2 z^2 + \ldots$$

ultimately behave like a geometric series, i.e. they satisfy

$$|a_k z^k| \le |a_k| r^k \le A$$

for some bound A, all large k and $|z| \le r$ for some $r > 0$. For $|z| < \frac{r}{2}$, this implies that the remainder for any n satisfies

$$\sum_{k=n+1}^{\infty} a_k z^k = O(z^{n+1}),$$

so the convergent power series for f for $|z| \le r$ is indeed asymptotic as $z \to 0$. More simply, recall Taylor series with remainder.

A general technique to obtain asymptotic expansions for integrals is again termwise integration. Consider, for example, the Laplace transform

$$I(x) = \int_0^\infty \frac{t^{\lambda-1}}{1+t} e^{-xt} \, dt \qquad (2.17)$$

for $\lambda > 0$ and x large. Since $\frac{1}{1+t} = \sum_{s=0}^{n-1}(-t)^s + \frac{(-t)^n}{1+t}$, we obtain

$$I(x) = \sum_{s=0}^{n-1} \frac{(-1)^s \Gamma(s+\lambda)}{x^{s+\lambda}} + r_n(x) \qquad (2.18)$$

in terms of Euler's *gamma (or factorial) function*

$$\Gamma(z) \equiv \int_0^\infty e^{-t} t^{z-1} \, dt, \quad \mathrm{Re}\ z > 0$$

(cf. Olver et al. [361]) for the remainder

$$r_n(x) \equiv (-1)^n \int_0^\infty \frac{t^{n+\lambda+1}}{1+t} e^{-xt} \, dt.$$

Since $|r_n(x)| \leq \frac{\Gamma(n+\lambda)}{x^{n+\lambda}}$,

$$I(x) \sim \sum_{s=0}^\infty \frac{(-1)^s \Gamma(s+\lambda)}{x^{s+\lambda}} \qquad \text{as } x \to \infty. \qquad (2.19)$$

Again, even though the Maclaurin series for $\frac{1}{1+t}$ only converges for $0 \leq t < 1$, its coefficients determine the asymptotics for $I(x)$ as $x \to \infty$. (Readers should understand such typical arguments.) Generalizations of this procedure to integrals

$$\int_0^\infty f(t) e^{-xt} \, dt$$

often are labeled *Laplace's method* (or *Watson's lemma*). A real variables approach to obtain such results is found in Olver [360], while a complex variables approach is presented in Wong [525]. More general techniques for the asymptotic evaluation of integrals include the *stationary phase* and *saddle point methods* (called Edgeworth expansions in statistics).

(c)   The WKB Method

The WKB method (cf. Olver [360], Schissel [436], Miller [318], Cheng [83], Wong [526], and Paulsen [387]) concerns asymptotic solutions of the scalar linear homogeneous second-order differential equation

$$y'' + \lambda^2 f(x, \lambda) y = 0 \qquad (2.20)$$

when the real parameter $\lambda \to \infty$ and f is bounded. Introducing the *logarithmic derivative*

$$u(x, \lambda) = \frac{y'(x, \lambda)}{y(x, \lambda)} = (\ln y)', \tag{2.21}$$

or equivalently setting

$$y = e^{\int^x u(s, \lambda)\, ds}, \tag{2.22}$$

converts the given linear second-order differential equation (2.20) to the non-linear first-order generalized Riccati equation

$$u' + u^2 + \lambda^2 f = 0 \tag{2.23}$$

(since $y' = uy$ and $y'' = (u' + u^2)y$) which, generally, can't be solved directly. (This is not the simple version solved by Count Riccati (or Johann Bernoulli) (cf. Roy [427]).) We will further suppose that the expansion

$$f(x, \lambda) \sim \sum_{n=0}^{\infty} \frac{f_n(x)}{\lambda^n} \tag{2.24}$$

is known and valid on an interval $\alpha < x < \beta$ as $\lambda \to \infty$. Then, we will seek a (formal) asymptotic solution

$$u(x, \lambda) \sim \lambda \sum_{n=0}^{\infty} \frac{u_n(x)}{\lambda^n}, \tag{2.25}$$

of (2.23) with corresponding series for u^2 and u'. Equating coefficients of λ^2 and λ^{2-n} in the differential equation, we will successively need

$$u_0^2 + f_0 = 0$$

and

$$u'_{n-1} + 2u_0 u_n + \sum_{k=1}^{n-1} u_k u_{n-k} + f_n = 0 \quad \text{for each} \quad n \geq 1.$$

Thus, we will take

$$u_0(x) = u_0^{\pm}(x) = \begin{cases} \pm i\sqrt{f_0(x)} & \text{if } f_0(x) > 0 \\ \pm \sqrt{-f_0(x)} & \text{if } f_0(x) < 0 \end{cases} \tag{2.26}$$

and

$$u_n(x) = -\frac{1}{2u_0(x)} \left(u'_{n-1} + \sum_{k=1}^{n-1} u_k u_{n-k} + f_n \right) \quad \text{for each} \quad n \geq 1. \tag{2.27}$$

In particular,

$$u_1(x) = -\frac{1}{2} \frac{d}{dx} \left(\ln u_0(x) \right) - \frac{f_1(x)}{2u_0(x)}$$

implies two linearly independent WKB approximates

$$y^{\pm}(x,\lambda) = \frac{1}{\sqrt[4]{f_0(x)}} e^{\pm i\lambda \int_{x_0}^{x} \left(\sqrt{(f_0(s))} - \frac{1}{2\lambda} \frac{f_1(s)}{\sqrt{f_0(s)}} \right) ds} (1 + o(1)) \quad \text{if } f_0(x) > 0$$

(2.28)

and

$$y^{\pm}(x,\lambda) = \frac{1}{\sqrt[4]{|f_0(x)|}} e^{\pm \lambda \int_{x_0}^{x} \left(\sqrt{|f_0(s)|} - \frac{1}{2\lambda} \frac{f_1(s)}{\sqrt{|f_0(s)|}} \right) ds} (1 + o(1)) \quad \text{if } f_0(x) < 0$$

(2.29)

for (2.20). The algebraic prefactor comes from the first term of u_1. See Keller and Lewis [245] for connections to geometrical optics and Keller [244] regarding the related Born and Rytov approximations. Note, further, that one consequence of the leading term approximation, important in quantum mechanics, is the so-called *adiabatic invariance* (cf. Arnold et al. [12] and Ou and Wong [385]). Knowing these linearly independent approximate solutions also allows us to solve the nonhomogeneous equation, i.e. to determine an asymptotic Green's function (cf. Stakgold [472]).

As defined above, the $o(1)$ symbol in (2.28–2.29) indicates an expression that goes to zero as $\lambda \to \infty$. Its approximate form would be determined by u_2. Miller [318] proves the validity of the WKB approximation using a contraction mapping argument, while Olver [360] bounds the error involved in terms of the total variation of a natural control function. Note the singularities of y that result at any turning points where f_0 has a zero. Also note that the solutions (2.28–2.29) change from being exponential to oscillatory (or vice-versa) as such points are crossed with f_0 changing signs.

As an alternative to (2.28), we could directly seek asymptotic solutions of (2.20) in the form

$$A(x,\lambda) e^{i\lambda \int^x \sqrt{f_0(s)}\, ds} + \overline{A(x,\lambda)} e^{-i\lambda \int^x \sqrt{f_0(s)}\, ds}$$

(2.30)

for a complex-valued asymptotic power series $A(x,\lambda)$ whose terms could be successively found using an undetermined coefficients scheme. Thus

$$y = A e^{i\lambda \int^x \sqrt{f_0(s)}\, ds} + \text{c.c.}$$

must satisfy the differential equation. Because

$$y'' = \left[A'' + 2i\lambda \sqrt{f_0(x)} A' + \frac{i\lambda}{2} \frac{f_0'(x)}{\sqrt{f_0(x)}} A - \lambda^2 f_0(x) A \right] e^{i\lambda \int^x \sqrt{f_0(s)}\, ds} + \text{c.c.},$$

we will need A to satisfy

$$\frac{1}{\lambda} \left[A'' + \left(\frac{f(x,\lambda) - f_0(x)}{\lambda} \right) A \right] + 2i\sqrt{f_0(x)} A' + \frac{i}{2} \frac{f_0'(x)}{\sqrt{f_0(x)}} A = 0 \quad (2.31)$$

as a power series

$$A\left(x, \frac{1}{\lambda}\right) \sim \sum_{j \geq 0} \frac{A_j(x)}{\lambda^j}.$$

Murray [338] works out a variety of WKB examples quite explicitly.

Olver points out that the separate results of the physicists Wentzel, Kramers, and Brillouin in 1926 and those of Jeffreys in 1924 were actually obtained independently by Joseph Liouville and George Green in 1837. Carlini had even treated a special case involving Bessel functions in 1817. See Heading [200] and Fröman and Fröman [164] for further history. Nonetheless, the WKB(J) label seems to persist. (William Thomson, later Lord Kelvin, visited Paris in 1845 after his Cambridge graduation and introduced Jacques Sturm and Liouville to the work [183] of Green, a recently deceased former miller from Nottingham, memorialized in 1993 with a plaque in Westminster Abbey near the tomb of Newton and plaques to Kelvin, Maxwell, and Faraday (cf. Cannell [69]). (Green's mill is now restored as a science center.) As late as 1953, Sir Harold Jeffreys called WKB

approximations of Green's type.

First, note that the WKB results provide existence and uniqueness theorems for the singularly perturbed linear ODE

$$\epsilon \frac{d^2 y}{dx^2} + a(x) \frac{dy}{dx} + b(x) y = 0 \tag{2.32}$$

when $\epsilon > 0$ is small, a and b are smooth, and $a(x) \neq 0$, when Dirichlet boundary conditions are applied at two endpoints, say α and β. Also note that the *Sturm transformation*

$$y(x) = w(x) e^{-\frac{1}{2\epsilon} \int^x a(s)\, ds}, \tag{2.33}$$

requires w to satisfy

$$\epsilon \frac{d^2 w}{dx^2} + f(x, \epsilon) w = 0 \tag{2.34}$$

for

$$f(x, \epsilon) \equiv b(x) - \frac{1}{2} a'(x) - \frac{1}{4\epsilon} a^2(x).$$

The transformation (2.33) holds for all ϵ, but we will be especially concerned with the more challenging situation that ϵ is small but positive. Multiplying (2.34) by w and integrating by parts, supposing homogeneous boundary conditions $w(\alpha) = w(\beta) = 0$, implies that

$$\epsilon \int_\alpha^\beta \left(\frac{dw}{ds}\right)^2 ds = \int_\alpha^\beta f(s, \epsilon)\, w^2(s)\, ds$$

since the boundary terms $\epsilon w \frac{dw}{dx}$ at α and β then vanish. Thus, $w(x) \equiv 0$ must hold when

$$f(x, \epsilon) \leq 0 \text{ on } \alpha \leq x \leq \beta, \tag{2.35}$$

and the given two-point problem for y then has a unique solution (just let w be the difference between any two of them). Note that the sign condition (2.35) on f is satisfied if either

(i) $a(x) \neq 0$ and $\epsilon > 0$ is small,

or

(ii) $2b \leq a'$ and $\epsilon > 0$.

Uniqueness conditions for more general Sturm-Liouville boundary value problems can be found in Courant-Hilbert [103] and Zettl [532].

Existence of the solutions y of (2.32) follows from using the two linearly independent real WKB solutions, which take the form

$$A(x, \epsilon) \quad \text{and} \quad B(x, \epsilon)e^{-\frac{1}{\epsilon}\int^x a(s)\, ds} \tag{2.36}$$

for asymptotic series A and B. (Note that this A is not the complex amplitude A used in the WKB solution (2.30).) In particular, the exponential factor in (2.33) is cancelled or doubled in the corresponding solutions (2.36).

The resulting *outer solution* $A(x, \epsilon)$ of (2.32) must satisfy

$$\epsilon A'' + a(x)A' + b(x)A = 0 \tag{2.37}$$

as a real power series in ϵ, so its leading term must satisfy

$$a(x)A_0' + b(x)A_0 = 0,$$

i.e.

$$A_0(x) = e^{-\int_{x_0}^x \frac{b(s)}{a(s)}\, ds} A_0(x_0).$$

Likewise $\left(Be^{\frac{1}{\epsilon}\int^x a(s)\, ds}\right)' = \left(B' - \frac{Ba}{\epsilon}\right)e^{-\frac{1}{\epsilon}\int^x a(s)\, ds}$ and $\left(Be^{-\frac{1}{\epsilon}\int^x a(s)\, ds}\right)'' = \left(B'' - \frac{2}{\epsilon}B'a - \frac{1}{\epsilon}Ba' + \frac{B}{\epsilon^2}a^2\right)e^{-\frac{1}{\epsilon}\int^x a(s)\, ds}$, so the differential equation for y requires that

$$\epsilon B'' - aB' + (b - a')B = 0. \tag{2.38}$$

Its leading term B_0 must satisfy $aB_0' + (a' - b)B_0 = 0$, so

$$B_0(x) = \frac{a(x_0)}{a(x)}e^{\int_{x_0}^x \frac{b(s)}{a(s)}\, ds} B_0(x_0)$$

and the general solution of (2.32) on $x \geq x_0$ takes the form

$$y(x, \epsilon) = e^{-\int_{x_0}^x \frac{b(s)}{a(s)}\, ds} A_0(x_0) + e^{-\frac{1}{\epsilon}\int_{x_0}^x \left(a(s) - \epsilon\frac{b(s)}{a(s)}\right)\, ds} B_0(x_0)\frac{a(x_0)}{a(x)} + O(\epsilon) \tag{2.39}$$

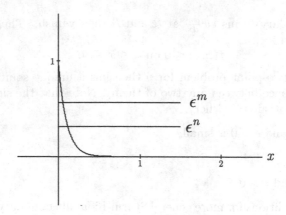

Figure 2.1: $e^{-x/\epsilon}$ for a small $\epsilon > 0$ is ultimately smaller than ϵ^κ for any $\kappa > 0$. Here, $n > m > 0$

when $a(x) > 0$. If *bounded* values $y(x_0)$ and $\epsilon y'(x_0)$ are prescribed, we will need

$$y(x_0) = A_0(x_0) + B_0(x_0) + O(\epsilon)$$

and

$$y'(x_0) = -\frac{1}{\epsilon}a(x_0)B_0(x_0) + O(1),$$

so

$$A_0(x_0) = y(x_0) - B_0(x_0) \tag{2.40}$$

and

$$B_0(x_0) = -\frac{\epsilon y'(x_0)}{a(x_0)}. \tag{2.41}$$

Having these two linearly independent solutions (2.36) to the linear differential equation (2.32) will allow us to asymptotically solve many boundary value problems for it and its nonhomogeneous analog. The errors made using such approximations are asymptotically negligible like $e^{-Cx/\epsilon}$ for some $C > 0$ and $x > 0$, so smaller than $O(\epsilon^n)$ for any $n > 0$. See Fig. 2.1 and Howls [219]. Note that the solution (2.39) will feature an initial layer of nonuniform convergence. The asymptotic solutions of (2.32) for $a(x) \neq 0$ will be more satisfactory throughout boundary layer regions than those traditionally found by matched expansions, as we shall later demonstrate. When such restrictions as $f(x, \epsilon) \leq 0$ in (2.34) don't hold, and for nonlinear generalizations, we must expect either multiple solutions to such two-point problems or none at all.

To illustrate typical behavior near a (simple) turning point, consider the equation

$$\epsilon^2 y'' - xh^2(x)y = 0 \tag{2.42}$$

for a smooth $h(x) > 0$. Oscillatory behavior for $x < 0$ and exponential behavior for $x > 0$ are provided by the WKB solutions. Locally, i.e. near the turning point $x = 0$, we naturally use the Airy equation

$$\ddot{w} - tw = 0 \qquad (2.43)$$

as a *comparison equation*. Its linearly independent solutions are the Airy functions $Ai(t)$, and $Bi(t)$ (cf. Olver et al. [361]). Their asymptotic behavior as $t \to \pm\infty$ is well known, e.g.,

$$Ai(t) \sim \frac{1}{2\sqrt{\pi}t^{1/4}} e^{-\frac{2}{3}t^{3/2}} \qquad \text{as} \quad t \to \infty$$
and
$$Ai(t) \sim \frac{1}{\sqrt{\pi}} \frac{1}{(-t)^{1/4}} \sin\left(\frac{2}{3}(-t)^{3/2} + \frac{\pi}{4}\right) \qquad \text{as} \quad t \to -\infty.$$

The connections to the WKB solutions follow the *Langer* transformation

$$y \sim \left(\frac{xh^2(x)}{S(x)}\right)^{1/4} \left[C_1 Ai\left(\frac{S(x)}{\epsilon^{2/3}}\right) + C_2 Bi\left(\frac{S(x)}{\epsilon^{2/3}}\right)\right] \qquad (2.44)$$

for constants C_1 and C_2,

$$S(x) = \left[\frac{3}{2}\int_0^x \sqrt{|r|}\, h(r)\, dr\right]^{2/3},$$

and the corresponding limits as $t = \frac{S(x)}{\epsilon^{2/3}} \to \pm\infty$ (cf. Wasow [515] for details of the so-called *connection problem*). Expansions can also be found for multiple and, even, coalescing turning points. Fowkes [152] solves the problem using multiple scale methods.

   More generally, it's often valuable to realize the equivalence of the Riccati differential equation

$$y' = a(x)y^2 + b(x)y + c(x) \qquad (2.45)$$

and second-order linear homogeneous differential equations. The transformation

$$y = \frac{-w'}{aw} \quad \text{or} \quad w = e^{-\int^x a(s)y(s)\, ds} \qquad (2.46)$$

in (2.45) implies that w will satisfy the linear equation

$$w'' - \left[\frac{a'(x)}{a(x)} + b(x)\right] w' + a(x)c(x)w = 0. \qquad (2.47)$$

Many times, its solutions can be provided in terms of special functions, thereby giving solutions y of the Riccati equation (2.45) as well through (2.46). On the other hand, if we can guess (or otherwise ascertain) a differentiable solution y of the Riccati equation, it determines a nontrivial solution w of the linear equation (2.47) and, by reduction of order, the general solution. Other transformations for linear equations are given in Kamke [232] and Fedoryuk [145].

Note, in particular, that the second-order linear equation

$$\epsilon u'' + b_1(x)u' + b_0(x)u = 0 \tag{2.48}$$

can be converted to

$$\epsilon q'' + (q')^2 + b_1(x)q' + \epsilon b_0(x) = 0 \tag{2.49}$$

by setting

$$u = e^{q/\epsilon} \tag{2.50}$$

and that the latter equation can be solved asymptotically by taking

$$q(x, \epsilon) \sim \sum_{j \geq 0} \epsilon^j q_j(x) \tag{2.51}$$

or, equivalently, by setting

$$u = C(x, \epsilon) e^{q_0(x)/\epsilon}$$

for a power series C (cf. Bender and Orszag [36]), which can be sought termwise with respect to ϵ.

(d) The Regular Perturbation Procedure

In the following, we shall consider it natural and straightforward (even central to singular perturbations) to use a *regular perturbation* method to find power series solutions to nonlinear vector initial value problems

$$\dot{x} = f(x, t, \epsilon), \quad t \geq 0, \quad x(0) = c(\epsilon) \tag{2.52}$$

based on knowing a smooth vector solution $x_0(t)$ to the limiting nonlinear problem

$$\dot{x}_0 = f(x_0, t, 0), \quad t \geq 0, \quad x_0(0) = c_0(0) \tag{2.53}$$

on some bounded interval $0 \leq t \leq T$. Assuming sufficient smoothness of f and c and the series expansions

$$f(x, t, \epsilon) \sim \sum_{j \geq 0} f_j(x, t)\epsilon^j$$

with smooth coefficients f_j and

$$c(\epsilon) \sim \sum_{j \geq 0} c_j \epsilon^j,$$

we shall let

$$x(t, \epsilon) \sim \sum_{k \geq 0} x_k(t)\epsilon^k. \tag{2.54}$$

Expanding about $\epsilon = 0$,

$$f(x(t,\epsilon),t,\epsilon) = f(x_0(t),t,0) + \left(f_x(x_0(t),t,0)(\epsilon x_1(t) + \epsilon^2 x_2(t) + \ldots)\right)$$
$$+ \epsilon f_\epsilon(x_0,t,0)) + \left(\frac{1}{2}((f_{xx}(x_0(t),t,0)))(\epsilon x_1(t) + \ldots)\right)(\epsilon x_1(t) + \ldots)$$
$$+ \epsilon f_{x\epsilon}(x_0(t),t,0)(\epsilon x_1(t) + \ldots) + \frac{\epsilon^2}{2}f_{\epsilon\epsilon}(x_0(t),t,0) + \ldots$$

and equating successive coefficients of powers of ϵ in (2.52), we naturally require

$$\dot{x}_1 = \frac{\partial f}{\partial x}(x_0(t),t,0)x_1 + \frac{\partial f}{\partial \epsilon}(x_0(t),t,0), \quad x_1(0) = c_1 \tag{2.55}$$

$$\dot{x}_2 = \frac{\partial f}{\partial x}(x_0(t),t,0)x_2 + \frac{1}{2}\left(\frac{\partial^2 f}{\partial x^2}(x_0(t),t,0)x_1(t)\right)x_1(t)$$
$$+ \frac{\partial^2 f}{\partial x \partial \epsilon}(x_0(t),t,0)x_1(t) + \frac{1}{2}\frac{\partial^2 f}{\partial \epsilon^2}(x_0(t),t,0)), \quad x_2(0) = c_2, \tag{2.56}$$

etc. These linear equations for x_j with $j \geq 1$ can be successively and uniquely solved using the nonsingular *fundamental matrix* Φ for the linearized homogeneous system

$$\dot{\Phi} = A(t)\Phi \quad \text{with} \quad \Phi(0) = I \tag{2.57}$$

for the Jacobian

$$A(t) = f_x(x_0(t),t,0)$$

and the identity matrix I (cf. Brauer and Nohel [59]). When A is constant, Φ is the matrix exponential e^{At}. Recall the variation of constants (parameters) method to solve the linear vector initial value problem

$$\dot{y} = A(t)y + b(t), \quad y(0) \text{ given.} \tag{2.58}$$

Set

$$y = \Phi(t)w(t)$$

for an unspecified vector w. First note that the unique Φ can be found by iterating in the integral equation

$$\Phi(t) = I + \int_0^t A(s)\Phi(s)\,ds$$

corresponding to (2.57). This yields the approximations

$$\Phi(t) = I + \int_0^t A(s)\,ds + \int_0^t A(s)\int_0^s A(r)\,dr\,ds + \ldots,$$

etc. (sometimes called the *matrizant*) which converge. Differentiating y, we get $\dot{y} = \dot{\Phi}w + \Phi\dot{w} = A\Phi w + b(t)$, so we will need

$$\Phi\dot{w} = b \quad \text{and} \quad y(0) = w(0).$$

Integrating, we uniquely obtain $w(t) = y(0) + \int_0^t \Phi^{-1}(s)b(s)\,ds$ (since Φ will be, at least locally, nonsingular). Thus,

$$y(t) = \Phi(t)y(0) + \int_0^t \Phi(t)\,\Phi^{-1}(s)\,b(s)\,ds, \qquad (2.59)$$

as can be readily checked.

Knowing the solution $x_0(t)$ (of (2.53)) for $\epsilon = 0$ and the resulting Φ (cf.(2.57)), we next obtain

$$x_1(t) = \Phi(t)c_1 + \int_0^t \Phi(t)\,\Phi^{-1}(s)\,f_\epsilon(x_0(s), s, 0)\,ds \qquad (2.60)$$

and, then, in turn

$$x_2(t) = \Phi(t)c_2 + \int_0^t \Phi(t)\,\Phi^{-1}(s)\left[\left(\frac{1}{2}f_{xx}(x_0(s), s, 0)\,x_1(s) + \right.\right.$$
$$\left.\left. f_{x\epsilon}(x_0(s), s, 0)\right)x_1(s) + \frac{1}{2}f_{\epsilon\epsilon}(x_0(s), s, 0)\right]ds, \qquad (2.61)$$

etc. We expect the series (2.54) for x to converge uniformly for ϵ small and bounded t. This justifies the regular perturbation technique, which we will henceforth apply routinely. When the assumptions don't apply, the asymptotic solution may not simply be a power series in ϵ. Puiseaux expansions in fractional powers of ϵ (cf. [17]) are a possibility. As we will ultimately find, however, we cannot expect to blindly use these approximate solutions on intervals where t becomes unbounded. A proof on finite intervals is given in de Jager and Jiang [224]. See Smith [466] for the celebrated example of Einstein's equation for the motion of Mercury about the sun.

More generally, one might also use such regular perturbation (i.e., power series) methods to solve operator equations

$$T(u, \epsilon) = 0 \qquad (2.62)$$

for small ϵ, justified by applying the *implicit function theorem* under appropriate conditions (cf. Miller [318]) to get an analytic solution $u(x, \epsilon)$ (cf. Baumgärtel [32] and Krantz and Parks [262]). Also, see Kato [243] and Avrachenkov et al. [17] regarding linear operators.

A classic example involves the zeroes of the Wilkinson polynomial

$$\prod_{k=1}^{20}(z - k) + \epsilon z^{19}$$

(cf. Wilkinson [523], Bender and Orszag [36], and Moler [324]). Its complex roots are extremely sensitive to perturbations. Corresponding to $k = 15$, the first correction is of the order $5\epsilon 10^{10}$, providing extreme sensitivity of the perturbation.

In ending the chapter, we want to emphasize that we have severely restricted the topics covered, keeping in mind our limited later needs. More generally, the use of iteration methods to obtain asymptotic expansions is often very efficient, as is the use of convergence acceleration methods (cf. Weniger [518]), among many other computational techniques. We recommend Barenblatt [27] and [28]'s unique development of *intermediate asymptotics*, relating and extending basic concepts from dimensional analysis, self-similarity, and scaling. Consulting the extensive literature cited is highly recommended! Bosley [53] even provides a numerical version of asymptotics.

Among recent texts, Zeytounian [533] attempts to model viscous, compressible, heat-conducting, Newtonian, baroclinic, and nonadiabatic fluid flow using

the art of modeling assisted, rationally, by the spirit of asymptotics.

Motivation for such *rational asymptotic modeling* is found in the autobiography Zeytounian [534].

Example 1

Gobbi and Spigler [171] consider the scalar singular linear two-point boundary value problem

$$\epsilon^2 u'' - u = -\frac{1}{\sqrt{x(1-x)}}, \quad 0 \le x \le 1, \quad u(0) = u(1) = 0. \quad (2.63)$$

Since the auxiliary polynomial $\epsilon^2 \lambda^2 - 1 = 0$ provides the complementary solutions $e^{-x/\epsilon}$ and $e^{-(1-x)/\epsilon}$, we can use variation of parameters to find the general solution

$$u(x, \epsilon) = e^{-x/\epsilon} C_1 + e^{-(1-x)/\epsilon} C_2 + \frac{1}{2\epsilon} \int_0^x \frac{e^{-(x-t)/\epsilon} \, dt}{\sqrt{t(1-t)}} + \frac{1}{2\epsilon} \int_x^1 \frac{e^{(x-t)/\epsilon} \, dt}{\sqrt{t(1-t)}}$$
$$(2.64)$$

of the nonhomogeneous differential equation (2.63). The boundary conditions imply that

$$C_1 \sim -\frac{1}{2\epsilon} \int_0^1 \frac{e^{-t/\epsilon} \, dt}{\sqrt{t(1-t)}} \quad \text{and} \quad C_2 \sim -\frac{1}{2\epsilon} \int_0^1 \frac{e^{-(1-t)/\epsilon} \, dt}{\sqrt{t(1-t)}}$$

up to asymptotically negligible amounts. Since $\Gamma(\frac{1}{2}) = \sqrt{\pi}$,

$$u(x, \epsilon) \sim -\frac{1}{2} \sqrt{\frac{\pi}{\epsilon}} \left(e^{-x/\epsilon} + e^{-(1-x)/\epsilon} \right)$$
$$+ \frac{1}{2\epsilon} \left[\int_0^x \frac{e^{-(x-t)/\epsilon} \, dt}{\sqrt{t(1-t)}} + \int_x^1 \frac{e^{(x-t)/\epsilon} \, dt}{\sqrt{t(1-t)}} \right]. \quad (2.65)$$

Within $(0,1)$, the first two terms are asymptotically negligible, while the primary contributions to the two integrals come from near $t = x$. Indeed,

$$\frac{1}{2\epsilon} \int_0^x \frac{e^{-(x-t)/\epsilon} \, dt}{\sqrt{t(1-t)}} \sim \frac{1}{2\sqrt{x(1-x)}}$$

and the second integral has the same limit. Thus, as expected,

$$u(x, \epsilon) \sim \frac{1}{\sqrt{x(1-x)}} \qquad \text{within} \quad (0,1). \tag{2.66}$$

Because the asymptotic solution is symmetric about $1/2$ and the outer limit is unbounded at the endpoints, further analysis is necessary to determine the asymptotic behavior in the endpoint boundary layers. Computations for $\epsilon = 0.01$ and 0.0025 provide spikes near 0 and 1 with $O(1/\sqrt{\epsilon})$ maxima, as found.

Example 2

Reiss [414] introduced the *combustion* model

$$\dot{y} = y^2(1 - y), \quad y(0) = \epsilon \tag{2.67}$$

(cf. Kapila [234]). Because $\dot{y} > 0$ for $0 < y < 1$, we know that the solution y increases monotonically from ϵ to the rest point 1 at $t = \infty$. The exact solution can be obtained by separating variables since integrating $\left(\frac{1}{y} + \frac{1}{1-y} + \frac{1}{y^2}\right) dy = dt$ implies that

$$\ln\left(\frac{y}{1-y}\right) - \frac{1}{y} = t - \frac{1}{\epsilon} + \ln\left(\frac{\epsilon}{1-\epsilon}\right). \tag{2.68}$$

This implicit result shows, e.g., that

$$y = \frac{1}{2} \quad \text{when} \quad t = \frac{1}{\epsilon} - \ln\left(\frac{\epsilon}{1-\epsilon}\right) - 2$$

while

$$y = \frac{9}{10} \quad \text{when } t = \frac{1}{\epsilon} - \ln\left(\frac{\epsilon}{1-\epsilon}\right) - \frac{10}{9} + \ln 9$$

and

$$y = \frac{99}{100} \quad \text{when } t = \frac{1}{\epsilon} - \ln\left(\frac{\epsilon}{1-\epsilon}\right) - \frac{100}{99} + \ln 99.$$

Thus, the ultimate explosion is long delayed when ϵ is small.

For bounded values of t, the *preignition* solution can be represented by a small regular perturbation expansion

$$Y(t, \epsilon) = \epsilon Y_1(t) + \epsilon^2 Y_2(t) + \dots$$

satisfying

$$\dot{Y}_1 = 0, \quad Y_1(0) = 1$$

and

$$\dot{Y}_2 = Y_1^2, \quad Y_2(0) = 0,$$

etc. termwise. Thus

$$Y(t, \epsilon) = \epsilon + \epsilon^2 l + \dots \tag{2.69}$$

This breaks down as the *slow-time* $\tau \equiv \epsilon t$ grows. Indeed, the *explosion* takes place about

$$\tilde{t} = \frac{1}{\epsilon} - \ln\left(\frac{\epsilon}{1-\epsilon}\right) - 2, \tag{2.70}$$

as can be verified numerically for, say, $\epsilon = 1/10$. One might say that a boundary layer (nonuniform convergence) occurs as $t \to \infty$.

Readers should be aware that one of the most successful texts presenting asymptotic methods has been Bender and Orszag [36], reprinted by Springer in 1999. Originated at MIT to teach the ubiquitous course in advanced mathematical methods for scientists and engineers, it featured easy, intermediate, difficult, and theoretical sections, corresponding exercises, and quotes from Sherlock Holmes at the beginning of each chapter. Paulsen [387] is a well-written new textbook seeking to simplify Bender and Orszag and make its subject more accessible.

To illustrate the centrality of asymptotics, we quote Dvortsina [128] regarding the prominent Soviet physicist I. M. Lifshitz (co-author of many outstanding texts with Nobel prizewinner Lev Landau):

> Everyone who knew Lifshitz remembers well that every time he began a discussion of any work he asked first of all: "What small parameter did you choose?" He meant to say that in the majority of problems solved by theoretical physics the smallness of some quantity is always used.

After reading this chapter, and perhaps trying the exercises, the author hopes you no longer think asymptotic approximations are some sort of mystical constructions. They're down to earth!

Exercises

1. (Awrejcewicz and Krysko [18]) Show that

$$\sin 2\epsilon \sim 2\epsilon - \frac{4}{3}\epsilon^3 + \frac{4}{15}\epsilon^5 + \dots$$
$$\sim 2\tan\epsilon - 2\tan^3\epsilon - 2\tan^5\epsilon + \dots$$

$$\sim 2\ln(1+\epsilon)+\ln(1+\epsilon^2)-2\ln(1+\epsilon^3)+\ln(1+\epsilon^4)+\frac{2}{5}\ln(1+\epsilon^5)+\dots$$

$$\sim 6\left(\frac{\epsilon}{3+2\epsilon^2}\right)-\frac{756}{5}\left(\frac{\epsilon}{3+2\epsilon^2}\right)^5+\dots.$$

2. (Awrejcewicz and Krysko [18]) Consider

$$f(x,\epsilon)=\sqrt{x+\epsilon}.$$

Note that $f(0,\epsilon)=\sqrt{\epsilon}$. Show that

$$f(x,\epsilon)=\sqrt{x}\left(1+\frac{\epsilon}{2x}-\frac{\epsilon^2}{8x^2}+\frac{\epsilon^3}{16x^3}-\dots\right)$$

when $\frac{\epsilon}{x}\to 0$, so the expansion is nonuniform.

3. (a) Find a power series expansion for the solution of the initial value problem

$$y'=1+y^2,\quad y(0)=\epsilon.$$

(b) Find the exact solution and determine the first four terms of its power series about $\epsilon=0$.

4. Find the first three terms of two power series solutions

$$u(x,\epsilon)=u_0(x)+\epsilon u_1(x)+\epsilon^2 u_2(x)+\dots$$

of the nonlinear differential equation

$$\epsilon u''=u^2-u+\epsilon x.$$

5. (a) Find a regular perturbation solution to the initial value problem

$$y'=1+y^2+\epsilon y,\quad y(0)=\epsilon.$$

Where does it become singular?

(b) Solve the equation

$$w''-\epsilon w'+1=0$$

and determine

$$y=-\frac{w'}{w}.$$

Where is it singular?

6. (cf. Hoppensteadt [213]) Consider the initial value problem

$$\dot{x}=-x^3+\epsilon x,\quad x(0)=1,\ t\geq 0.$$

(a) Find the first two terms of the regular perturbation expansion.

(b) Find the exact solution and its limit as $t \to \infty$.

(c) Explain the breakdown of the regular perturbation expansion.

7. (Hsieh and Sibuya [220]) Consider the two-point problem

$$y'' = \epsilon \sin\left(\frac{x}{100 - y^2}\right), \quad -1 \le x \le 1, \quad y(-1) = y(1) = 0.$$

Obtain the solution $y = \phi(x, c, \epsilon)$ by "shooting," i.e. by solving the initial value problem

$$y'' = \epsilon \sin\left(\frac{x}{100 - y^2}\right), \quad y(-1) = 0, \quad y'(-1) = c(\epsilon)$$

for an appropriate $c(\epsilon)$. Determine the first two terms in the power series for $y(x, \epsilon)$ and $c(\epsilon)$. Observe the extensive and effective use of the shooting method in Hastings and McLeod [198].

8. A typical ODE exercise is to compute the terms of the power series solution to the initial value problem

$$y' = x^2 + y^2, \quad y(0) = 2.$$

(a) Set

$$y(x) = \sum_{n=0}^{\infty} C_n x^n$$

and show that $C_0 = 2$, $C_1 = 4$, $C_2 = 8$, $C_3 = \frac{49}{3}$, and $C_{n+1} = \frac{1}{n+1} \sum_{p=0}^{n} C_p C_{n-p}$, $n \ge 3$.

(b) Convert the equation to the Weber equation

$$w'' + x^2 w = 0$$

by using the representation $w = e^{-\int_0^x y(s)\, ds}$.

(c) Find the power series for w about $x = 0$, checking that $y = -\frac{w'}{w}$. The radius of convergence for w is infinite. Note that w can be expressed in terms of parabolic cylinder (or Weber) functions.

9. (Kevorkian and Cole [249]) Consider the initial value problem

$$u'' + u = \epsilon f(x) u$$

$$u(0) = 0, \quad u'(0) = 1, \quad x \ge 0.$$

Show that a necessary condition that the regular perturbation expansion of the solution be uniformly valid on $x \ge 0$ is to have $\int_0^x f(s)\, ds$ bounded there.

Chapter 3

The Method of Matched Asymptotic Expansions and Its Generalizations

(a) Elementary Matching

Milton Van Dyke's *Perturbation Methods in Fluid Mechanics* [490] was effectively both the earliest and the most influential book specifically about applied singular perturbations. (Some credit might be given earlier fluid dynamics textbooks, e.g., Hayes and Probstein [199]). Van Dyke extensively surveyed the large extant aeronautical and fluid dynamical literature, forcefully advocating and clarifying the so-called *method of matched asymptotic* (or *inner* and *outer*) *expansions*. Although Van Dyke acknowledged that Prandtl's boundary layer theory was the prototype singular perturbation problem, he introduced the subject by describing incompressible fluid flow past a thin airfoil. The book's highlight message, sometimes called Van Dyke's *magic rule*, states:

> The m-term inner expansion of (the n term outer expansion) = the n-term outer expansion of (the m term inner expansion).

This glib oversimplification (for any positive integer pairs m and n) allowed many practitioners to confidently solve significant applied problems asymptotically (an advantage unavailable before then).

© Springer International Publishing Switzerland 2014   53
R. O'Malley, *Historical Developments in Singular Perturbations*,
DOI 10.1007/978-3-319-11924-3_3

To grasp the basic idea of Van Dyke's procedure for $m = n = 2$, consider the linear initial value problem

$$\begin{cases} \epsilon y'' + y' + y = 0 & \text{on } 0 \le t \le T \quad \text{for a fixed finite time } T \\ y(0) = 0, \ y'(0) = \frac{1}{\epsilon} & \text{for a small } \epsilon > 0 \end{cases} \qquad (3.1)$$

for a displacement y. We expect the impulsive large initial derivative to provide an immediate rapid upward response, so we naturally introduce the *fast time*

$$\tau = t/\epsilon. \qquad (3.2)$$

Then $y' = \frac{1}{\epsilon} y_\tau$ and $\epsilon y'' = \frac{1}{\epsilon} y_{\tau\tau}$, so we naturally seek a local *inner* expansion y^{in} satisfying the *stretched* problem

$$\begin{cases} y_{\tau\tau} + y_\tau + \epsilon y = 0 \text{ on } \tau \ge 0 \\ \text{with } y(0) = 0 \text{ and } y_\tau(0) = 1. \end{cases} \qquad (3.3)$$

(Sophisticated readers will note that our selection of the *stretched variable* τ rebalances the orders of the three terms in the given ODE, changing their *dominant balance* in the terminology of Bender and Orszag [36]. To determine the "right" balance will more generally take some trial and error. The selection of the stretched variable also relates to a classical asymptotic technique called the *Newton polygon* (cf. Hille [205], Kung and Traub [269], and White [519]) which is implemented in MAPLE. Setting

$$y^{in}(\tau, \epsilon) \sim y_0(\tau) + \epsilon y_1(\tau) + \epsilon^2 y_2(\tau) + \ldots \qquad (3.4)$$

and expanding y_τ and $y_{\tau\tau}$ analogously, we will need to satisfy

$$(y_{0\tau\tau} + \epsilon y_{1\tau\tau} + \ldots) + (y_{0\tau} + \epsilon y_{1\tau} + \ldots) + \epsilon(y_0 + \ldots) = 0,$$

or

$$y_{0\tau\tau} + y_{0\tau} + \epsilon(y_{1\tau\tau} + y_{1\tau} + y_0) + \ldots = 0,$$

and the corresponding initial conditions

$$y(0, \epsilon) = y_0(0) + \epsilon y_1(0) + \ldots = 0$$

and

$$y_\tau(0, \epsilon) = y_{0\tau}(0) + \epsilon y_{1\tau}(0) + \ldots = 1$$

as a regular perturbation expansion in powers of ϵ. Equating coefficients, we naturally require y_0 to satisfy

$$y_{0\tau\tau} + y_{0\tau} = 0, \quad y_0(0) = 0, \quad \text{and} \quad y_{0\tau}(0) = 1, \qquad (3.5)$$

and y_1 to next satisfy

$$y_{1\tau\tau} + y_{1\tau} + y_0 = 0, \quad y_1(0) = y_{1\tau}(0) = 0, \qquad (3.6)$$

etc. Thus, we uniquely obtain

$$y_0(\tau) = 1 - e^{-\tau} \tag{3.7}$$

while $y_{1\tau\tau} + y_{1\tau} + 1 - e^{-\tau} = 0$ and the trivial initial conditions uniquely imply that

$$y_1(\tau) = 2 - \tau + e^{-\tau}(-2 - \tau) \tag{3.8}$$

(using, say, the method of undetermined coefficients).

We then expect the resulting uniquely determined series

$$y^{in}(\tau, \epsilon) = 1 - e^{-\tau} + \epsilon(2 - \tau - e^{-\tau}(2 + \tau)) + \ldots \tag{3.9}$$

or *inner expansion* to be asymptotically valid at least for bounded τ values, i.e. for small values of $t = O(\epsilon)$. (It breaks down when τ is large, since the *ratio of successive terms* in the series ultimately becomes unbounded like $\epsilon\tau$.)

For larger values of t, we shall alternatively seek an *outer* solution Y^{out}, depending on the original time variable t and ϵ. Thus, we will substitute the regular power series (i.e., *outer*) expansion

$$Y^{out}(t, \epsilon) \sim Y_0(t) + \epsilon Y_1(t) + \ldots \tag{3.10}$$

into the given differential equation and equate coefficients of like powers of ϵ in (3.1) to successively require $Y_0' + Y_0 = 0$, $Y_1' + Y_1 + Y_0'' = 0$, etc. Hence

$$Y_0(t) = Ae^{-t} \quad \text{for some constant } A, \tag{3.11}$$

$$Y_1(t) = (B - At)e^{-t} \quad \text{for some constant } B, \tag{3.12}$$

etc., providing the first terms of an outer expansion

$$Y^{out}(t, \epsilon) = Ae^{-t} + \epsilon(B - At)e^{-t} + \ldots \tag{3.13}$$

for finite t values and constants A, B, ... yet to be determined by *matching* this *outer expansion* to the inner expansion (3.9) as we now describe. (Note that the terms Y_k in (3.10) satisfy first, rather than second, order differential equations and that the prescribed initial values at $t = 0$ are so far irrelevant to the outer expansion.) In the 1950s, an alternative *patching* technique was sometimes applied to inner and outer expansions. Patching typically took place at an ϵ-dependent t value like $-10\epsilon \ln \epsilon$. The concept still underlies some numerical methods (cf., e.g., Kopteva and O'Riordan [259] and Miller et al. [317] regarding the *Shishkin mesh*).

We first rewrite the known inner expansion in terms of the outer variable t as

$$y^{in}\left(\frac{t}{\epsilon}, \epsilon\right) = 1 - e^{-t/\epsilon} + \epsilon\left(2 - \frac{t}{\epsilon} - e^{-t/\epsilon}\left(2 + \frac{t}{\epsilon}\right)\right) + \ldots$$

Taking the limit as $\tau = t/\epsilon \to \infty$, the exponentials become negligible and we get the truncated two-term limit

$$(y^{in})^{out} = 1 - t + 2\epsilon + \ldots \tag{3.14}$$

Analogously, we represent the outer expansion in terms of the inner variable τ as

$$Y^{out}(\epsilon\tau, \epsilon) = Ae^{-\epsilon\tau} + \epsilon(B - \epsilon\tau A)e^{-\epsilon\tau} + \dots$$

Expanding the exponentials in their Maclaurin expansions for moderate values of τ as $\epsilon \to 0$ and truncating, we obtain

$$(Y^{out})^{in} \equiv A + \epsilon(B - A\tau) + \dots. \tag{3.15}$$

Since $\tau = \epsilon t$, the *asymptotic matching condition*

$$(y^{in})^{out} = (Y^{out})^{in}$$

(at this ($m = n = 2$) order) requires that

$$
\begin{aligned}
(y^{in})^{out} &= 1 - t + 2\epsilon + \dots \\
&= A - At + \epsilon B + \dots = (Y^{out})^{in}.
\end{aligned}
\tag{3.16}
$$

We will naturally call this expression the *common part* of the inner and outer expansions (both truncated at the second order). We could express it in terms of either time variable t or τ. Note that the matching condition crudely corresponds to the idea of equating $Y^{out}(t, \epsilon)$ near $t = 0$ to $y^{in}(\tau, \epsilon)$ near $\tau = \infty$. We are, however, being much more explicit.

This process uniquely provides the unspecified constants $A = 1$ and $B = 2$ in the outer expansion, i.e., matching across the $O(\epsilon)$-thick initial layer (by equating the common parts) has uniquely specified the outer expansion as

$$Y^{out}(t, \epsilon) = e^{-t} + \epsilon(2 - t)e^{-t} + \dots \tag{3.17}$$

We expect (3.17) to be the valid asymptotic solution for t *outside* the initial layer. Note that $Y^{out}(0, 0) \neq y(0)$. (If this wasn't so, the inner and outer expansions would coincide for $t = \epsilon\tau$.) Note that we seem to implicitly invoke some idea about *overlap* of the two solutions in a joint region of validity of the inner and outer expansions.

Rather than having separate asymptotic expansions, y^{in} very near $t = 0$ and Y^{out} away from $t = 0$, we shall now define the additive *composite* expansion

$$
\begin{aligned}
y^c &\equiv Y^{out} + y^{in} - (Y^{out})^{in} \\
&= (e^{-t} + \epsilon(2 - t)e^{-t} + \dots) \\
&\quad + (1 - e^{-\tau} + \epsilon[2 - \tau - (2 + \tau)e^{-\tau}] + \dots) \\
&\quad - (1 - t + 2\epsilon + \dots) \\
&= [e^{-t} - e^{-\tau}] + \epsilon[(2 - t)e^{-t} - (2 + \tau)e^{-\tau}] + \dots
\end{aligned}
\tag{3.18}
$$

that we expect to be *uniformly valid* on any fixed finite interval $0 \leq t \leq T$ as $\epsilon \to 0$, i.e. in the domains where the inner expansion, the outer expansion, and their common part are simultaneously defined. The limit of the sum y^c is y^{in} in the inner region and Y^{out} in the outer region since the outer expansion in the inner region and the inner expansion in the outer region agree with their common (i.e., *matching*) part. (We note that other alternative composite expansions have also been introduced in the literature.) Eckhaus [133] formalizes the procedure using *expansion operators*. Van Dyke [492] noted that the terminology *global* and *local* approximations would be preferable to outer and inner approximations.

A more subtle matching technique using *intermediate variables*

$$t_\beta \equiv \frac{t}{\epsilon^\beta}$$

for βs satisfying $0 < \beta < 1$ in both the inner and outer expansions is presented in Cole [92] and Holmes [209]. For some problems, the use of power series in ϵ for both the inner and outer expansions turns out to be inadequate for matching, but inserting intermediate terms suggested by their limits succeeds. The process is called *switchback*. To avoid going wrong, Van Dyke [491] made the practical suggestion

Don't cut between logarithms.

Its subtle meaning could be clarified by examining detailed examples that caused anxiety 40 years ago.

The exact solution to the initial value problem (3.1) has the form

$$y(t, \epsilon) = C(\epsilon)(e^{-\nu(\epsilon)t} - e^{-\kappa(\epsilon)t/\epsilon}) \tag{3.19}$$

where

$$\nu(\epsilon) \equiv \frac{1 - \sqrt{1 - 4\epsilon}}{2\epsilon} \sim 1 + \epsilon + \dots$$

and

$$\kappa(\epsilon) \equiv \frac{1 + \sqrt{1 - 4\epsilon}}{2} \sim 1 - \epsilon - \epsilon^2 + \dots.$$

Thus, $y(0) = 0$ and $y'(0) = \frac{1}{\epsilon} = C(\epsilon)\left(-\nu(\epsilon) + \frac{\kappa(\epsilon)}{\epsilon}\right)$ uniquely determine

$$C(\epsilon) \equiv \frac{1}{\kappa(\epsilon) - \epsilon\nu(\epsilon)} = \frac{1}{1 - 2\epsilon + \dots} = 1 + 2\epsilon + \dots.$$

The exact result

$$y(t, \epsilon) \sim \frac{1}{1 - 2\epsilon + \dots}\left(e^{-(1+\epsilon+\dots)t} - e^{-(1-\epsilon-\epsilon^2+\dots)\frac{t}{\epsilon}}\right) \tag{3.20}$$

agrees asymptotically with the composite solution (3.18) obtained by matching for $m = 2$ and $n = 2$. (To carry out these calculations, we use the *binomial expansion* $\sqrt{1-x} = 1 - \frac{x}{2} - \frac{x^2}{8} + \ldots$, convergent for $|x| < 1$.) Readers should personally experiment by matching solutions of (3.1) for larger values of m and n than 2.

Further, as we will extensively illustrate, matching results in the same uniform expansion as we'd get by determining the *outer expansion* $Y(t, \epsilon)$ as a function of t (with its unspecified constants) and adding to it a *boundary layer corrector* expansion $v(\tau, \epsilon)$ (as a function of the stretched time $\tau = t/\epsilon$) that tends exponentially to zero as $\tau \to \infty$. Thus, we'll have

$$y(t, \epsilon) \sim Y(t, \epsilon) + v(\tau, \epsilon) \tag{3.21}$$

Matching, then, ultimately cancels some terms in the inner expansion (retaining $v \equiv y^{in} - (y^{in})^{out}$), but it is somewhat inefficient because it requires us to determine terms in y^{in} that are later neglected (i.e., the common part).

Specifically, note that the exact solution (3.20) of (3.1) also has the form

$$y(t, \epsilon) = Y(t, \epsilon) + Z(t, \epsilon)e^{-t/\epsilon} \tag{3.22}$$

for power series Y and Z depending on t and ϵ. Indeed, for bounded t, $Y \equiv C(\epsilon)e^{-\nu(\epsilon)t}$ is the outer solution. The initial conditions require that

$$y(0) = Y(0, \epsilon) + Z(0, \epsilon) = 0$$

and

$$\epsilon y'(0) \equiv \epsilon Y'(0, \epsilon) + \epsilon Z'(0, \epsilon) - Z(0, \epsilon) = 1.$$

Since y, Y and the *corrector* $v \equiv Ze^{-t/\epsilon}$ all satisfy the given differential equation of (3.1), Z must then satisfy

$$\epsilon Z'' - Z' + Z = 0 \tag{3.23}$$

as a series in ϵ. The representation (3.22) implies a more efficient power series method than matching. More sophisticated matching procedures for linear differential equations in the complex plane are considered in Olde Daalhuis et al. [359]. Likewise, the Russian A. M. I'lin [221] convincingly presents matching for partial differential equations.

The unusual problem

$$(x + \epsilon)y' + y = 0, \quad y(1) = 1$$

has the exact solution

$$y(x, \epsilon) = \frac{1 + \epsilon}{x + \epsilon},$$

well-behaved for $0 < x \leq 1$, but algebraically unbounded near $x = 0$ where the limiting equation has a singular point. Complications there must be expected (cf. our discussion of Lighthill's method in Chap. 5).

Exercises

1. Show that $e^{-t/\epsilon} \leq \epsilon^n$ holds for $-n\epsilon \ln \epsilon \leq t < \infty$ and that the inequality is reversed for smaller $t > 0$.

2. For the initial value problem

$$\epsilon \ddot{y} + \dot{y} + y = 0, \quad t \geq 0, \quad y(0) = 1, \quad \dot{y}(0) = 1,$$

   show that the asymptotic solution has the form

$$y(t, \epsilon) = Y(t, \epsilon) + \epsilon D(t, \epsilon)e^{-t/\epsilon}$$

   on $0 \leq t \leq T < \infty$ for power series Y and D. The uniform limit for $t \geq 0$ will be $Y_0(t) = e^{-t}$, since $\dot{Y}_0 + Y_0 = 0$ and $Y_0(0) = 1$. Show that $\dot{y}$ will jump near $t = 0$, however. Try computing the solution and its derivative for a small ϵ.

3. (Cole [92]) The equation

$$\epsilon y'' + (1 + \alpha x)y' + \alpha y = 0$$

   is exact, so it is possible to obtain its general solution. Suppose $\alpha > -1$, so $1 + \alpha x > 0$ on $0 \leq x \leq 1$. Impose the boundary values

$$y(0) = 0 \text{ and } y(1) = 1,$$

   so the outer limit is $Y_0(x) = \frac{1+\alpha}{1+\alpha x}$. Note that the limiting initial layer corrector

$$-(1 + \alpha)e^{-x/\epsilon}$$

   approximates the exact corrector

$$-(1 + \alpha)e^{-\frac{1}{\epsilon}\int_0^x (1+\alpha s)\,ds}.$$

   Find the exact solution and the first two terms of its outer solution

$$Y(x, \epsilon) = Y_0(x) + \epsilon Y_1(x) + O(\epsilon^2).$$

4. Consider the alternative composite expansion y^c for problem (3.1) when the common part is nonzero by setting

$$y^c = \frac{Y^{out}y^{in}}{((Y^{out})^{in})^2}.$$

5. Consider the two-point problem

$$\epsilon y'' + (1 + x)^2 y' + 2(1 + x)y = 0, \quad 0 < x < 1$$

$$y(0) = 1, \quad y(1) = 2.$$

(a) Obtain the exact solution and describe its limiting behavior. (Hint: the differential equation is exact.)

(b) Determine a composite expansion in the form

$$y(x, \epsilon) = A(x, \epsilon) + v(\xi, \epsilon)$$

where A is an outer expansion valid for $x > 0$ and the boundary layer corrector $v \to 0$ as $\xi \equiv x/\epsilon \to \infty$.

(c) Determine an asymptotic solution in the WKB form

$$y(x, \epsilon) = A(x, \epsilon) + e^{-\frac{1}{\epsilon} \int_0^x (1+s)^2 \, ds}(y(0) - A(0, \epsilon)).$$

6. Consider the two-point problem

$$\epsilon y'' + (1+x)^2 y' - (1+x)y = 0, \quad 0 \le x \le 1$$

with $y(0) = 1$ and $y(1) = 3$.

(a) Obtain the exact solution and determine its limiting behavior as $\epsilon \to 0^+$. (Hint: $y = 1 + x$ is a solution of the ODE.)

(b) Use matched asymptotic expansions to obtain the two-term composite expansion.

(c) Determine an asymptotic solution of the form

$$y(x, \epsilon) = A(x, \epsilon) + B(x, \epsilon)e^{-\frac{1}{\epsilon} \int_0^x (1+s)^2 \, ds}$$

(with power series expansions for A and B).

(d) Plot the inner expansion, the outer expansion, the composite expansion, and the numerical solution for $\epsilon = 1/10$ (on the same graph).

(e) Show that

$$e^{-\frac{1}{\epsilon} \int_0^x (1+s)^2 \, ds} - e^{-\frac{x}{\epsilon}} = O(\epsilon) \quad \text{on} \quad 0 \le x \le 1.$$

7. Assuming a boundary layer of $O(\epsilon)$ thickness near $x = 1$, seek an asymptotic solution of

$$\epsilon u_{xx} = u_x + u_t, \quad u(0, t) = u_0(t), \quad u(1, t) = u_1(t) \quad \text{for } t \ge 0$$
$$\text{and } u(x, 0) \text{ given for } 0 \le x \le 1$$

in the form

$$u(x, t, \epsilon) = A(x, t, \epsilon) + B(x, t, \epsilon)e^{-(1-x)/\epsilon}.$$

Basic issues concerning the validity of matching were raised by Fraenkel [155] and Eckhaus [133], among others (cf., e.g., Lo [297] and, especially, Skinner [463]). Some of the subtleties were reconsidered in the annotated edition of Van Dyke's book [491] of 1975. Its frontispiece is the woodcut *Sky and Water I, 1938* by the Dutch lithographer M. C. *Escher* featuring fish transforming vertically into birds (cf. Schattschneider [433] and [434] regarding relations between Escher's work and groups, tilings, and other mathematical objects). (The author recently found this print for sale for about $48,000!) Van Dyke stated that the woodcut

> gives a graphical impression of the "imperceptively smooth blending" of one flow into another that is the heart of the method of matched asymptotic expansions.

Milton Van Dyke (1922–2010) was an American who got a 1949 Caltech Ph.D. (with Paco Lagerstrom) and worked at NASA-Ames before taking a professorship in aeronautics at Stanford in 1959 (see Schwartz [442] for a brief biography). One reason for the annotated edition [491] of *Perturbation Methods in Fluid Mechanics* was that Academic Press let the 1964 original [490] go out of print because Van Dyke had insisted that the contract stipulate that

> the book shall cost no more than three cents a page.

The Academic Press edition sold 8,000 copies. (In addition to the annotated edition, Parabolic Press (managed by Van Dyke) also published the picture book *An Album of Fluid Motion* (1984) by Van Dyke and the autobiographical *Stories of a 20th Century Life* (1994) by W. R. Sears.)

The more complicated use of intermediate limits/intermediate problems, rather than the formal matching of series, as proposed by Kaplun [235], relates to the often *presumed* existence of an *overlap* (as in analytic continuation in complex variables) between the domains of validity for the inner and outer expansions and the construction of a "composite" or uniform expansion as the formal sum of the inner and outer expansions less their "common part," found by matching. Eckhaus and Fraenkel both showed that having an overlap is not necessary for matching to succeed. Fruchard and Schäfke [165], however, base their *composite* expansions on overlap. (The complication that the inner and outer expansions are expressed in terms of different variables indeed suggests the more sophisticated *two-timing* (or *multiple scale*) procedure that we will consider in Chap. 5.) The recent proofs of Skinner [463] and Fruchard and Schäfke [165] validate matching for a broad variety of ODE problems.

Fluid dynamicists have introduced a more elaborate *triple deck* technique (cf. Meyer [316], Sobey [467], and Veldman [498], noting important contributions by Stewartson, Williams, and Neiland) to handle viscous flow along a plate. Somewhat analogously, mathematicians have introduced a *blow-up* technique to analyze even more complicated matching (cf. Dumortier and Roussarie [127] and Kuehn [268]).   Hastings and McLeod [198] combine blowup with classical methods to rigorously prove matching for the Lagerstrom model

$$y'' + \frac{n-1}{r}y' + \epsilon y y' = 0, \quad r \geq 1, \quad y(1) = 0, \quad y(\infty) = 1$$

in dimensions $n = 2$ and 3. This problem has been considered by a dozen authors since 1957. Most recently, Holzer and Kaper [211] used normal form techniques to handle a variety of problems with so-called *logarithmic switchback*.

(b)   Tikhonov–Levinson Theory and Boundary Layer Corrections

(i)   Introduction

Wolfgang Wasow's *Asymptotic Expansions for Ordinary Differential Equations* [513] is a much more mathematical work than Van Dyke [490]. It is centered on singular perturbations, but also includes the study of regular and irregular singular points, as well as turning points. Much of the theory is carried out using matrix differential equations (which may have limited its appeal to the very applied audience). Its singular perturbation coverage includes boundary value problems for linear scalar ordinary differential equations, following Wasow's [517] NYU doctoral thesis, as well as the (perhaps less efficient) methods of the prominent Russian analysts Vishik and Lyusternik [507, 508] and Pontryagin [398]. Results for nonlinear initial value problems rely on papers by the Soviet academician Andrei Nikolaevich Tikhonov (1906–1993) on the solution of

> systems of equations with a small parameter in the term with the highest derivative

(a large percentage of singular perturbation problems, as we shall find). Tikhonov's work on asymptotics appeared from 1948 to 1952 and was continued in the ongoing work of his former student Adelaida B. Vasil'eva (1926–) (Ph.D., Moscow State, 1961) (cf. Vasil'eva, Butuzov, and Kalachev [496] and earlier monographs in Russian by Vasil'eva and by she and her former student and MSU colleague Vladimir Butuzov). Instead of matching per se, she directly obtains a composite expansion by the so-called *boundary function method*, a technique analogous to the *boundary layer correction* method or "the subtraction trick" (which first finds the outer solution (formally) and then subtracts it from the solution being sought. Matching is then simple because the new outer expansion and the new common part are both trivial) (cf. Lions [295], O'Malley [366, 368], Smith [466], or Verhulst [500]). For a survey of Soviet work, see Vasil'eva [495].

We point out that J.-L. Lions (1928–2001) led a large school of French analysts (including many prominent former students) who applied asymptotics to control, stochastic, and partial differential equations. Readers are encouraged to consult their publications, e.g., [295].

The basic Tikhonov results were largely independently obtained later by Norman Levinson (1912–1975), senior author of the long-dominant ODE textbook Coddington and Levinson [91]. Levinson's approach was more geometric, aimed at describing *relaxation oscillations*, as occur for the van der Pol equation (cf. Levinson [287]), anticipating much recent work involving invariant manifolds. Related work was done with his junior colleague Earl Coddington and by a number of MIT graduate students from the 1950s, including D. Aronson, R. Davis, L. Flatto, V. Haas, S. Haber, J. Levin, V. Mizel, R. O'Brien, J. Scott-Thomas, and D. Trumpler.

(ii)   A Nonlinear Example

To get an idea of Tikhonov–Levinson theory, we will first consider the specific planar initial value example

$$\begin{cases} \dot{x} = y, & x(0) = 1 \\ \epsilon\dot{y} = x^2 - y^2, & y(0) = 0 \end{cases} \tag{3.24}$$

on a bounded $t \geq 0$ interval as $\epsilon \to 0^+$ (or the equivalent initial value problem for the second-order nonlinear scalar equation $\epsilon\ddot{x} + (\dot{x})^2 - x^2 = 0$), followed by the linear vector system

$$\begin{cases} \dot{x} = A(t)x + B(t)y \\ \epsilon\dot{y} = C(t)x + D(t)y \end{cases}$$

and then the nonlinear system

$$\begin{cases} \dot{x} = f(x, y, t, \epsilon) \\ \epsilon\dot{y} = g(x, y, t, \epsilon) \end{cases}$$

with appropriate smoothness and stability assumptions. We shall characterize the system dynamics for (3.24) as being *slow-fast*, with variable x being slow compared to y (since the velocity $\dot{y} = O(1/\epsilon)$ when $x^2 \neq y^2$ while $\dot{x} = O(1)$ for bounded y). The *reduced* problem (obtained for $\epsilon = 0$)

$$\begin{cases} \dot{X}_0 = Y_0, & X_0(0) = 1 \\ 0 = X_0^2 - Y_0^2 \end{cases} \tag{3.25}$$

omits the initial condition for y and implies the two possible roots

$$Y_0 = \pm X_0 \tag{3.26}$$

of the algebraic equation, so X_0 must satisfy either initial value problem

$$\dot{X}_0 = \pm X_0, \quad X_0(0) = 1. \tag{3.27}$$

Hence, possible outer limits for $t > 0$ are

$$(X_0(t), Y_0(t)) = (e^{\pm t}, \pm e^{\pm t}). \tag{3.28}$$

Because $Y_0(0) = \pm 1$, while $y(0) = 0$, the fast variable y must initially converge nonuniformly. This suggests that we might actually have uniformly valid limits

$$\begin{cases} x(t, \epsilon) \sim X_0(t) \\ \text{and} \\ y(t, \epsilon) \sim Y_0(t) + v_0(\tau) \end{cases} \tag{3.29}$$

for bounded ts, where $v_0(0) = y(0) - Y_0(0)$ is the initial jump in the fast variable and the initial layer corrector v_0 is significant only in a thin initial layer where $v_0 \to 0$ as the fast time $\tau = \frac{t}{\epsilon}$ ranges from 0 to ∞. Thus, the correction term $v_0(\tau)$ provides the needed nonuniform convergence of the coordinate y in the $O(\epsilon)$-thick initial layer near $t = 0$, described in terms of τ. Then, we need

$$\epsilon \frac{dy}{dt} \sim \epsilon \frac{dY_0}{dt} + \frac{dv_0}{d\tau} \sim X_0^2 - (Y_0 + v_0)^2.$$

Since $Y_0 = \pm X_0 = \pm e^{\pm t}$ is bounded (for t bounded), $\epsilon \frac{dy}{dt} \sim \frac{dv_0}{d\tau}$ shows that v_0 must nearly satisfy $\frac{dv_0}{d\tau} \sim -2Y_0(\epsilon \tau)v_0 - v_0^2$. If we choose $Y_0(t) = e^t$, Y_0 will be nearly 1 near $t = 0$, so v_0 must satisfy the initial value problem

$$\frac{dv_0}{d\tau} = -(2 + v_0)v_0, \quad v_0(0) = -1 \tag{3.30}$$

on $\tau \geq 0$. This problem is easy to solve explicitly as a Riccati equation. Indeed, checking the sign of $\frac{dv_0}{d\tau}$ shows that v_0 increases monotonically from -1 to 0 as τ goes from 0 to ∞. We shall say that the initial vector $\begin{pmatrix} x(0) \\ y(0) \end{pmatrix} = \begin{pmatrix} 1 \\ 0 \end{pmatrix}$ lies in the *domain of influence* (or "region of attraction") of the root $Y_0 = X_0$ of the reduced problem (3.25). If we, instead, tried using the other possible root $Y_0 = -X_0 = -e^{-t}$, the corresponding v_0 would have to satisfy

$$\frac{dv_0}{d\tau} \sim v_0(2 - v_0), \quad v_0(1) = 1,$$

but then $v_0 \to 2$ as $\tau \to \infty$ would contradict the asymptotic stability required for the limiting initial layer correction v_0. That one root of the limiting equation (3.25) is *repulsive* and thereby inappropriate corresponds to our expectation that there be a unique asymptotic solution to the given initial value problem (3.24).

Vasil'eva's work (as well as O'Malley's) further suggests that the asymptotic solution of our initial value problem (3.24) indeed has the (higher-order) *composite form*

$$\begin{cases} x(t,\epsilon) = X(t,\epsilon) + \epsilon u(\tau,\epsilon) \\ y(t,\epsilon) = Y(t,\epsilon) + v(\tau,\epsilon) \end{cases} \tag{3.31}$$

uniformly on fixed bounded intervals $0 \le t \le T$, where the *outer solution* $\begin{pmatrix} X(t,\epsilon) \\ Y(t,\epsilon) \end{pmatrix}$ has an asymptotic power series expansion

$$\begin{pmatrix} X(t,\epsilon) \\ Y(t,\epsilon) \end{pmatrix} \sim \sum_{j \ge 0} \begin{pmatrix} X_j(t) \\ Y_j(t) \end{pmatrix} \epsilon^j \tag{3.32}$$

with

$$\begin{pmatrix} X_0(t) \\ Y_0(t) \end{pmatrix} = \begin{pmatrix} 1 \\ 1 \end{pmatrix} e^t \tag{3.33}$$

and where all terms of the scaled supplemental *initial layer corrector*

$$\begin{pmatrix} u(\tau,\epsilon) \\ v(\tau,\epsilon) \end{pmatrix} \sim \sum_{j \ge 0} \begin{pmatrix} u_j(\tau) \\ v_j(\tau) \end{pmatrix} \epsilon^j \tag{3.34}$$

in (3.31) tend to zero as the fast time

$$\tau = t/\epsilon \tag{3.35}$$

tends to infinity. Nonuniform convergence in the fast variable y (through v) provokes nonuniform convergence in the derivative $\dot{x}$ of the slow variable since $y = \dot{x} = Y + v$. That is why X (as compared to Y) has the asymptotically less significant initial layer correction ϵu. (Although we indicate full asymptotic expansions in (3.32) and (3.34), we in practice only generate a few terms of all the series.) The critical point is that our ansatz (3.31), especially its stability condition, usually allows us to bypass the tedium and inefficiency of actually matching inner and outer expansions. (A possible exception arises in singular cases when the outer limit $\begin{pmatrix} X_0(t) \\ Y_0(t) \end{pmatrix}$ is no longer defined or smooth at the initial point $t = 0$.)

Away from $t = 0$, the outer solution $\begin{pmatrix} X \\ Y \end{pmatrix}$ must satisfy the given system

$$\begin{cases} \dot{X} = Y \\ \epsilon \dot{Y} = X^2 - Y^2 \end{cases} \tag{3.36}$$

as a power series (3.32) in ϵ, since the initial layer correction $\begin{pmatrix} \epsilon u \\ v \end{pmatrix}$ and its derivative have decayed to zero there. For $\epsilon = 0$, we get the reduced system, and we pick its unique *attractive* solution $\begin{pmatrix} X_0 \\ Y_0 \end{pmatrix} = \begin{pmatrix} 1 \\ 1 \end{pmatrix} e^t$ because the other possibility did not allow the needed asymptotic stability of $v_0(\tau)$. From the coefficient of ϵ in (3.36), we require that

$$\begin{cases} \dot{X}_1 = Y_1 \\ \dot{Y}_0 = 2X_0X_1 - 2Y_0Y_1. \end{cases}$$

Since $\dot{Y}_0 = e^t = 2e^t(X_1 - Y_1)$, we need $\dot{X}_1 = Y_1 = X_1 - \frac{1}{2}$, so we obtain

$$X_1(t) = e^t\left(X_1(0) - \frac{1}{2}\right) + \frac{1}{2} \tag{3.37}$$

for an unspecified value $X_1(0)$. Higher-order terms $\begin{pmatrix} X_k \\ Y_k \end{pmatrix}$ in the outer expansion likewise also follow readily and uniquely, up to specification of the initial values $X_k(0)$ for each $k > 0$.

Returning to the slow equation $\dot{x} = y$, $\dot{x} = \dot{X} + \frac{du}{d\tau} = Y + v$, so $\dot{X} = Y$ implies the linear initial layer equation

$$\frac{du}{d\tau} = v. \tag{3.38}$$

Since $\epsilon \dot{Y} = X^2 - Y^2$, the nonlinear fast equation $\epsilon \dot{y} = \epsilon \dot{Y} + \frac{dv}{d\tau} = (X + \epsilon u)^2 - (Y + v)^2$ implies the coupled initial layer equation

$$\frac{dv}{d\tau} = -2Y(\epsilon\tau, \epsilon)v - v^2 + 2\epsilon X(\epsilon\tau, \epsilon)u + \epsilon^2 u^2, \tag{3.39}$$

with the terms of the outer solution $\begin{pmatrix} X \\ Y \end{pmatrix}$ already known, up to specification of $X(0, \epsilon)$. The initial conditions

$$\begin{cases} 1 = X(0, \epsilon) + \epsilon u(0, \epsilon) \\ 0 = Y(0, \epsilon) + v(0, \epsilon) \end{cases} \tag{3.40}$$

indeed termwise imply that

$$\begin{cases} 1 = X_0(0) \\ 0 = Y_0(0) + v_0(0) \end{cases} \tag{3.41}$$

and

$$\begin{cases} 0 = X_k(0) + u_{k-1}(0) \\ 0 = Y_k(0) + v_k(0) \end{cases} \tag{3.42}$$

for each $k \geq 1$. Thus, (3.41) requires

$$v_0(0) = -Y_0(0) = -1, \tag{3.43}$$

while (3.42) successively determines the unknown

$$X_k(0) = -u_{k-1}(0) \tag{3.44}$$

and, thereby, both $Y_k(0)$ (from the outer problem for X_k) and then $v_k(0)$.
Thus, the limiting initial layer system

$$\begin{cases} \frac{du_0}{d\tau} = v_0 \\ \frac{dv_0}{d\tau} = -2Y_0(0)v_0 - v_0^2 = -v_0(2 + v_0) \end{cases} \tag{3.45}$$

for (3.38–3.39) is subject to the initial condition $v_0(0) = -1$. A direct integration provides

$$v_0(\tau) = \tanh \tau - 1, \tag{3.46}$$

leaving the terminal value problem $\frac{du_0}{d\tau} = \tanh \tau - 1$, $u_0(\infty) = 0$. Integrating backwards from $\tau = \infty$, we uniquely get

$$u_0(\tau) = \ln \cosh \tau - \tau + \ln 2. \tag{3.47}$$

This immediately provides the needed initial value

$$X_1(0) = -u_0(0) = -\ln 2, \tag{3.48}$$

which uniquely specifies the second term $\begin{pmatrix} X_1 \\ Y_1 \end{pmatrix}$ of the outer solution via
(3.37). In particular, $Y_1(0) = X_1(0) - \frac{1}{2}$ next specifies

$$v_1(0) = -Y_1(0) = \frac{1}{2} - \ln 2, \tag{3.49}$$

by (3.42), while v_1 by (3.39) must satisfy the linear differential equation

$$\frac{dv_1}{d\tau} = -2(Y_0(0) + v_0(\tau))v_1 - 2(\tau Y_0'(0) + Y_1(0))v_0 + 2X_0(0)u_0. \tag{3.50}$$

Integrating this linear initial value problem provides $v_1(\tau)$ explicitly (though we won't bother to write down its expression) and the uniform approximations

$$x(t, \epsilon) = e^t + \epsilon \left[\frac{1}{2} - \left(\frac{1}{2} + \ln 2 \right) e^t + \ln \left(\cosh \frac{t}{\epsilon} \right) - \frac{t}{\epsilon} + \ln 2 \right] + O(\epsilon^2) \tag{3.51}$$

and

$$y(t, \epsilon) = e^t + \tanh \frac{t}{\epsilon} - 1 + \epsilon \left[\left(\frac{1}{2} + \ln 2 \right) (1 - e^t) + v_1 \left(\frac{t}{\epsilon} \right) \right] + O(\epsilon^2). \tag{3.52}$$

The blowup of e^t as $t \to \infty$ suggests that the results only apply on bounded t intervals. Hoppensteadt [212] added the necessary hypothesis that the solution of the reduced problem be asymptotically stable to Tikhonov's original conditions in order to extend the Tikhonov–Levinson theory to the infinite t interval. Also see Vasil'eva [494], however. Before proceeding, the reader should note (with some amazement) the efficient interlacing construction of the expansions for the outer solution and the initial layer correction. Readers should also observe how closely Tikhonov–Levinson theory links singular perturbations and stability theory (cf. Cesari [74] and Coppel [96]).

(iii)   Linear Systems

For the linear vector system

$$\begin{cases} \dot{x} = A(t)x + B(t)y \\ \epsilon\dot{y} = C(t)x + D(t)y \end{cases} \tag{3.53}$$

of $m + n$ scalar equations on, say, $0 \le t \le 1$, with smooth coefficients and prescribed bounded initial vectors

$$x(0) \quad \text{and} \quad y(0), \tag{3.54}$$

we will again seek a composite asymptotic solution of the form

$$\begin{aligned} x(t, \epsilon) &= X(t, \epsilon) + \epsilon u(\tau, \epsilon) \\ y(t, \epsilon) &= Y(t, \epsilon) + v(\tau, \epsilon) \end{aligned} \tag{3.55}$$

for $\tau = \epsilon t$, presuming the $n \times n$ matrix

$$D(t) \quad \text{remains } strictly\ stable \tag{3.56}$$

(i.e., has all its eigenvalues strictly in the left half-plane) for $0 \le t \le 1$.

Here, the limiting outer solution $\begin{pmatrix} X_0(t) \\ Y_0(t) \end{pmatrix}$ must satisfy the reduced problem

$$\begin{cases} \dot{X}_0 = A(t)X_0 + B(t)Y_0, \quad X_0(0) = x(0) \\ 0 = C(t)X_0 + D(t)Y_0. \end{cases}$$

Thus,

$$Y_0(t) = -D^{-1}(t)C(t)X_0(t) \tag{3.57}$$

and X_0 must be found as a solution of the reduced initial value problem

$$\dot{X}_0 = (A(t) - B(t)D^{-1}(t)C(t))X_0, \quad X_0(0) = x(0) \tag{3.58}$$

of m equations. Note that the state matrix for X_0 in (3.58) is the *Schur complement* of the block D in the matrix $\begin{pmatrix} A & B \\ C & D \end{pmatrix}$. Higher-order terms

$\begin{pmatrix} X_k \\ Y_k \end{pmatrix}$ in the outer expansion are determined from a regular perturbation solution of the system

$$\begin{cases} \dot{X} = A(t)X + B(t)Y \\ \epsilon \dot{Y} = C(t)X + D(t)Y \end{cases} \tag{3.59}$$

about $\begin{pmatrix} X_0 \\ Y_0 \end{pmatrix}$, i.e. from the nonhomogeneous system

$$\begin{aligned} \dot{X}_j &= (A(t) - B(t)D^{-1}(t)C(t))X_j + B(t)D^{-1}(t)\dot{Y}_{j-1} \\ Y_j &= -D^{-1}(t)C(t)X_j + D^{-1}(t)\dot{Y}_{j-1}. \end{aligned} \tag{3.60}$$

Moreover, linearity and the representation (3.55) imply that

$$\frac{dx}{dt} = \frac{dX}{dt} + \frac{du}{d\tau} \quad \text{and} \quad \epsilon\frac{dy}{dt} = \epsilon\frac{dY}{dt} + \frac{dv}{d\tau},$$

so the initial layer correction must satisfy the nearly constant coefficient system

$$\begin{cases} \frac{du}{d\tau} = \epsilon A(\epsilon\tau)u + B(\epsilon\tau)v \\ \frac{dv}{d\tau} = \epsilon C(\epsilon\tau)u + D(\epsilon\tau)v \end{cases} \tag{3.61}$$

and the limiting initial layer correction $\begin{pmatrix} u_0 \\ v_0 \end{pmatrix}$ must satisfy

$$\begin{cases} \frac{du_0}{d\tau} = B(0)v_0 \\ \frac{dv_0}{d\tau} = D(0)v_0, \quad v_0(0) = y(0) - Y_0(0). \end{cases} \tag{3.62}$$

Integrating, we explicitly obtain the decaying n-vector

$$v_0(\tau) = e^{D(0)\tau}(y(0) + D^{-1}(0)C(0)x(0)) \tag{3.63}$$

while

$$u_0(\tau) = -B(0)\int_\tau^\infty v_0(s)\, ds = B(0)D^{-1}(0)v_0(\tau). \tag{3.64}$$

Those unfamiliar with the matrix exponential should consult, e.g., Bellman [35].

Next, u_1 and v_1 will be decaying solutions of the initial value problem

$$\begin{cases} \frac{du_1}{d\tau} = B(0)u_1 + \tau\dot{B}(0)v_0 + A(0)u_0 \\ \frac{dv_1}{d\tau} = D(0)v_1 + \tau\dot{D}(0)v_0 + C(0)u_0, \quad v_1(0) = -Y_1(0) \end{cases}$$

which can be directly and uniquely solved. Taken vectorwise, the representation (3.55) determines the asymptotics of all solutions, i.e. of a fundamental matrix (cf. Coppel [96]) for the linear system (3.53) featuring initial layer behavior near $t = 0$.

(iv)   Nonlinear Systems

The ansatz (3.55) used further applies directly to the initial value problem for the general slow-fast nonlinear system

$$\begin{cases} \dot{x} = f(x,y,t,\epsilon), & x(0) \text{ given} \\ \epsilon\dot{y} = g(x,y,t,\epsilon), & y(0) \text{ given} \end{cases} \tag{3.65}$$

of $m+n$ smooth differential equations on $t \geq 0$ when the limiting differential-algebraic system (or reduced problem)

$$\begin{cases} \dot{X}_0 = f(X_0, Y_0, t, 0), & X_0(0) = x(0) \\ 0 = g(X_0, Y_0, t, 0) \end{cases} \tag{3.66}$$

(with m differential equations) has a smooth isolated solution

$$Y_0 = \phi(X_0, t) \tag{3.67}$$

of the n algebraic constraint equations $g = 0$ selected so that

(i) the resulting initial value problem

$$\dot{X}_0 = f(X_0, \phi(X_0, t), t, 0), \quad X_0(0) = x(0) \tag{3.68}$$

has a solution $X_0(t)$ defined on a finite interval $0 \leq t \leq T$ such that the Jacobian

$$g_y(X_0, \phi(X_0, t), t, 0)$$

remains a *stable* $n \times n$ matrix there and

(ii) the corresponding n-vector

$$v(0) = y(0) - \phi(x(0), 0)$$

lies in the domain of influence of the trivial solution of the limiting autonomous initial layer system

$$\frac{dv}{d\tau} = g(x(0), \phi(x(0), 0) + v, 0, 0) \quad \text{for } \tau = t/\epsilon \geq 0. \tag{3.69}$$

Hypothesis (i) provides *stability* for the outer solution $\begin{pmatrix} X \\ Y \end{pmatrix}$ on the finite t interval (and, via the implicit function theorem, guarantees that the root ϕ is locally unique), while hypothesis (ii) provides *asymptotic stability* for v as $\tau \to \infty$ within the initial layer, allowing the termwise construction of a decaying initial layer correction $\begin{pmatrix} \epsilon u \\ v \end{pmatrix}$ for $\tau \geq 0$. As an alternative, one could express condition (ii) in terms of the existence of an appropriate *Liapunov function* (cf. Khalil [251]). In practice, we begin by checking the hypotheses for various roots Y_0 of

$$g(X_0(t), Y_0(t), t, 0) = 0.$$

Smooth ϵ-dependent initial values for (3.65) would pose no complication.

We have naturally presumed that outside any initial layers the limiting solution to any singularly perturbed initial value problem satisfies the reduced problem, but this isn't always so. Eckhaus [133] introduced the counterexample

$$\epsilon^3 \cos\left(\frac{t}{\epsilon^2}\right) \ddot{x} + \epsilon \sin\left(\frac{t}{\epsilon^2}\right) \dot{x} - x = 0, \ t \geq 0$$

with initial values $x(0) = 1$ and $\dot{x}(0) = 0$. Its solution,

$$x(t, \epsilon) = 1 + \epsilon - \epsilon \cos\left(\frac{t}{\epsilon^2}\right),$$

however, tends to 1, rather than 0, as $\epsilon \to 0$.

We further note that one practical way to approximate the solution of a *differential-algebraic system*

$$\begin{cases} \dot{x} = f(x, y, t) \\ 0 = g(x, y, t) \end{cases}$$

is to *regularize* it, i.e. to introduce its singular perturbation

$$\begin{cases} \dot{x} = f(x, y, t) \\ \epsilon \dot{y} = g(x, y, t) \end{cases}$$

and to approximately solve that for a small positive ϵ (cf. O'Malley and Kalachev [374] and Nipp and Stoffer [351]).

As anticipated, we shall seek an asymptotic solution

$$\begin{pmatrix} x(t, \epsilon) \\ y(t, \epsilon) \end{pmatrix} = \begin{pmatrix} X(t.\epsilon) \\ Y(t, \epsilon) \end{pmatrix} + \begin{pmatrix} \epsilon u(\tau, \epsilon) \\ v(\tau, \epsilon) \end{pmatrix} \tag{3.70}$$

to (3.65) where the vector initial layer correction $\begin{pmatrix} \epsilon u \\ v \end{pmatrix} \to 0$ as $\tau \to \infty$.

(Vasil'eva, typically, does not introduce the ϵ multiplying u in the x-variable representation of (3.70). After some effort, however, she gets a trivial leading term for u.) Thus, the outer solution $\begin{pmatrix} X \\ Y \end{pmatrix}$ must satisfy the given system

$$\begin{cases} \dot{X} = f(X, Y, t, \epsilon) \\ \epsilon \dot{Y} = g(X, Y, t, \epsilon) \end{cases} \tag{3.71}$$

as a power series $\begin{pmatrix} X \\ Y \end{pmatrix} \sim \sum_{j \geq 0} \begin{pmatrix} X_j(t) \\ Y_j(t) \end{pmatrix} \epsilon^j$ in ϵ.

Further, the outer limit
$$\begin{pmatrix} X_0(t) \\ Y_0(t) \end{pmatrix}$$

must correspond to an *attractive* root $Y_0 = \phi$ of the limiting algebraic equation of (3.66) such that $g_y(X_0, Y_0, t, 0)$ is stable and the resulting initial value problem

$$\dot{X}_0 = f(X_0(t), \phi(X_0(t), t), t, 0), X_0(0) = x(0) \tag{3.72}$$

for the m-vector X_0 is guaranteed solvable (at least locally) by the classical existence and uniqueness theorem. Later terms $\begin{pmatrix} X_j \\ Y_j \end{pmatrix}$ must satisfy linearized systems

$$\dot{X}_j = f_x(X_0, Y_0, t, 0)X_j + f_y(X_0, Y_0, t, 0)Y_j + f_{j-1}(t) \tag{3.73}$$
$$0 = g_x(X_0, Y_0, t, 0)X_j + g_y(X_0, Y_0, t, 0)Y_j + g_{j-1}(t) \tag{3.74}$$

for $j > 0$, where f_{j-1} and g_{j-1} are known successively in terms of preceding coefficients. We obtain Y_j as an affine function of X_j from (3.74) because the Jacobian $g_y(X_0, Y_0, t, 0)$ remains nonsingular. This leaves a linear system for X_j, from (3.73), which will be uniquely solved once its initial value $X_j(0)$ is specified. Because $\dot{x} = \dot{X} + \frac{du}{d\tau}$, while $\epsilon \dot{y} = \epsilon \dot{Y} + \frac{dv}{d\tau}$, the initial layer correction $\begin{pmatrix} \epsilon u \\ v \end{pmatrix}$ must satisfy the nonlinear system

$$\begin{cases} \frac{du}{d\tau} = f(X + \epsilon u, Y + v, \epsilon\tau, \epsilon) - f(X, Y, \epsilon\tau, \epsilon) \\ \frac{dv}{d\tau} = g(X + \epsilon u, Y + v, \epsilon\tau, \epsilon) - g(X, Y, \epsilon\tau, \epsilon) \end{cases} \tag{3.75}$$

as a power series
$$\begin{pmatrix} u(\tau, \epsilon) \\ v(\tau, \epsilon) \end{pmatrix} \sim \sum_{j \geq 0} \begin{pmatrix} u_j(\tau) \\ v_j(\tau) \end{pmatrix} \epsilon^j$$

in ϵ. The t-dependent coefficients in (3.75) are expanded as functions of τ. Thus, $\begin{pmatrix} u_0 \\ v_0 \end{pmatrix}$ must satisfy

$$\frac{du_0}{d\tau} = f(X_0(0), Y_0(0) + v_0, 0, 0) - f(X_0(0), Y_0(0), 0, 0) \tag{3.76}$$

and

$$\frac{dv_0}{d\tau} = g(X_0(0), Y_0(0) + v_0, 0, 0) - g(X_0(0), Y_0(0), 0, 0) \tag{3.77}$$
$$= g(x(0), \phi(x(0), 0) + v_0, 0, 0).$$

Since

$$v_0(0) = y(0) - Y_0(0) = y(0) - \phi(x(0), 0)$$

has been assumed in hypothesis (ii) to lie in the domain of influence of the rest point $v_0 = 0$ of system (3.77), we are guaranteed that the nonlinear initial value problem for v_0 has the desired decaying solution $v_0(\tau)$ on $\tau \geq 0$. (One might need to obtain it numerically.) In terms of it, we simply integrate (3.76) to get

$$u_0(\tau) = -\int_\tau^\infty [f(x(0), \phi(x(0), 0) + v_0(s), 0, 0) \\ - f(x(0), \phi(x(0), 0), 0, 0)]\, ds. \tag{3.78}$$

This, in turn, provides the initial value

$$X_1(0) = -u_0(0) \tag{3.79}$$

needed to specify the outer expansion term $X_1(t)$ and thereby $Y_1(t)$. Linearized problems for $\begin{pmatrix} u_j \\ v_j \end{pmatrix}$, $j > 0$, with successively determined initial vectors $v_j(0) = -Y_j(0)$ will again have exponentially decaying solutions. This immediately specifies the needed vector $X_{j+1}(0) = -u_j(0)$ for the next terms in the outer expansion.

In the unusual situation that the outer solution

$$\begin{pmatrix} X(t, \epsilon) \\ Y(t, \epsilon) \end{pmatrix}$$

satisfies the initial condition

$$\begin{pmatrix} X(0, \epsilon) \\ Y(0, \epsilon) \end{pmatrix} = \begin{pmatrix} x(0) \\ y(0) \end{pmatrix},$$

the resulting boundary layer correction

$$\begin{pmatrix} \epsilon u(\tau, \epsilon) \\ v(\tau, \epsilon) \end{pmatrix}$$

will be trivial. If we have

$$y(0) = \phi(x(0), 0, 0),$$

we can omit the trivial first terms $\begin{pmatrix} u_0 \\ v_0 \end{pmatrix}$ of the boundary layer correction. Later terms naturally satisfy linear problems.

A substantial simplification occurs when the nonlinear system (3.71) is linear with respect to the fast variable y. Thus, we separately consider the initial value problem for the system

$$\begin{cases} \dot{x} = A(x, t, \epsilon) + B(x, t, \epsilon)y \\ \epsilon\dot{y} = C(x, t, \epsilon) + D(x, t, \epsilon)y \end{cases} \tag{3.80}$$

on $t \geq 0$. The corresponding reduced system

$$\begin{cases} \dot{X}_0 = A(X_0, t, 0) + B(X_0, t, 0)Y_0 \\ 0 = C(X_0, t, 0) + D(X_0, t, 0)Y_0 \end{cases} \tag{3.81}$$

will imply

$$Y_0(t) = -D^{-1}(X_0, t, 0)C(X_0, t, 0)$$

while X_0 must satisfy the reduced nonlinear vector problem

$$\dot{X}_0 = A(X_0, t, 0) - B(X_0, t, 0)D^{-1}(X_0, t, 0)C(X_0, t, 0), \ X_0(0) = x(0). \tag{3.82}$$

We suppose that (3.82) has a solution $X_0(t)$ on $0 \leq t \leq T < \infty$ with a resulting *stable* matrix

$$D(X_0(t), t, 0).$$

Higher-order terms in the outer expansion $\begin{pmatrix} X(t, \epsilon) \\ Y(t, \epsilon) \end{pmatrix}$ then follow successively, without complication, up to specification of $X(0, \epsilon)$.

The supplemental initial layer correction

$$\begin{pmatrix} \epsilon u(\tau, \epsilon) \\ v(\tau, \epsilon) \end{pmatrix} \tag{3.83}$$

must be a decaying solution of the stretched system

$$\begin{cases} \frac{du}{d\tau} = A(X + \epsilon u, t, \epsilon) - A(X, t, \epsilon) \\ \qquad + B(X + \epsilon u, t, \epsilon)(Y + v) - B(X, t, \epsilon)Y \\ \frac{dv}{d\tau} = C(X + \epsilon u, t, \epsilon) - C(X, t, \epsilon) \\ \qquad + D(X + \epsilon u, t, \epsilon)(Y + v) - D(X, t, \epsilon)Y. \end{cases} \tag{3.84}$$

Moreover, the initial conditions require

$$\begin{cases} X(0, \epsilon) + \epsilon u(0, \epsilon) = x(0) \\ \text{and} \\ Y(0, \epsilon) + v(0, \epsilon) = y(0). \end{cases} \tag{3.85}$$

Thus, the limiting linear initial layer problem is

$$\begin{cases} \frac{du_0}{d\tau} = B(x(0), 0, 0)v_0 \\ \frac{dv_0}{d\tau} = D(x(0), 0, 0)v_0, \quad v_0(0) = y(0) - Y_0(0). \end{cases} \tag{3.86}$$

It has the decaying solution

$$v_0(\tau) = e^{D(x(0), 0, 0)\tau} \left(y(0) + D^{-1}(x(0), 0, 0)C(x(0), 0, 0) \right) \tag{3.87}$$

with

$$u_0(\tau) = -\int_\tau^\infty B(x(0), 0, 0)v_0(s)\, ds$$
$$= -B(x(0), 0, 0)D^{-1}(x(0), 0, 0)v_0(\tau). \tag{3.88}$$

Later terms follow readily. Again, $X_1(0) = -u_0(0)$ will specify the $\mathcal{O}(\epsilon)$ terms in the outer expansion.

A special case of (3.65) is provided by the scalar *Liénard* equation

$$\epsilon\ddot{x} + f(x)\dot{x} + g(x) = 0 \tag{3.89}$$

on $t \geq 0$ with initial values $x(0)$ and $\dot{x}(0)$ provided. We introduce $y = \dot{x}$, so $\epsilon\dot{y} + f(x)y + g(x) = 0$. Then, the limiting solution is the monotonic solution of the separable equation

$$\dot{X}_0 = -\frac{g(X_0)}{f(X_0)}, \qquad X_0(0) = x(0), \tag{3.90}$$

presuming the stability hypothesis

$$f(X_0) < 0 \tag{3.91}$$

holds throughout. Then

$$y(t, \epsilon) = \dot{x}(t, \epsilon) = -\frac{g(X_0)}{f(X_0)} + e^{f(x(0))\tau}\left(\dot{x}(0) + \frac{g(x(0))}{f(x(0))}\right) + O(\epsilon) \tag{3.92}$$

features an initial layer while there is an implicit solution for $X_0(t)$.

Skinner [463] considers linear turning point problems of the special form

$$\epsilon^2 y' + xa(x, \epsilon)y = \epsilon b(x, \epsilon), \quad y(0) = \epsilon\alpha(\epsilon) \tag{3.93}$$

for smooth functions a and b with $a(x, \epsilon) > 0$. The simplest example seems to be

$$\epsilon^2 y' + xy = \epsilon, \quad y(0) = 0. \tag{3.94}$$

Its exact solution is the inner solution

$$y(x, \epsilon) = u(x/\epsilon) \tag{3.95}$$

for

$$u(\xi) = e^{-\frac{\xi^2}{2}}\int_0^\xi e^{\frac{r^2}{2}}\, dr. \tag{3.96}$$

Integrating by parts repeatedly, we get the algebraically decaying behavior

$$e^{-\frac{\xi^2}{2}}\int^\xi \frac{1}{r}\frac{d}{dr}\left(e^{\frac{r^2}{2}}\right)\, dr \sim \frac{1}{\xi} + \frac{1}{\xi^3} + \dots \qquad \text{as } \xi \to \infty,$$

corresponding to the readily generated outer expansion

$$Y(x,\epsilon) \sim \frac{\epsilon}{x} + \frac{\epsilon^3}{x^3} + \dots, \tag{3.97}$$

singular at the turning point $x = 0$. These problems are certainly more complicated than the initial value problems we have considered previously, so Skinner [463] is highly recommended reading.

Exercises

1. Find the exact solution to the scalar equation

$$\epsilon \dot{y} = y - y^3$$

   on $t \geq 0$ and determine how the outer limit $Y_0(t)$ for $t > 0$ depends on $y(0)$.

2. Solve the initial value problem for the planar system

$$\begin{cases} \dot{x} = xy \\ \epsilon \dot{y} = y - y^3 \end{cases}$$

   on $0 \leq t \leq T < \infty$ and determine the outer solution.

3. Obtain an $O(\epsilon^2)$ approximation to the solution of the planar initial value problem

$$\begin{cases} \dot{x} = -x + (x + \kappa - \lambda)y, & x(0) = -1 \\ \epsilon \dot{y} = x - (x + \kappa)y, & y(0) = 0 \end{cases}$$

   for positive constants κ and λ as $\epsilon \to 0^+$. The problem arises in enzyme kinetics (cf. Segel and Slemrod [448], Murray [339], and Segel and Edelstein-Keshet [447]).

4. A model for *autocatalysis* is given by the slow-fast system

$$\begin{cases} \dot{x} = x(1 + y^2) - y, & x(0) = 1 \\ \epsilon \dot{y} = -x(1 + y^2) + e^{-t}, & y(0) = 1. \end{cases}$$

   Seek an asymptotic solution of the form

$$x(t,\epsilon) = X(t,\epsilon) + \epsilon u(\tau,\epsilon)$$
$$y(t,\epsilon) = Y(t,\epsilon) + v(\tau,\epsilon)$$

   where $\begin{pmatrix} u \\ v \end{pmatrix} \to 0$ as $\tau = \frac{t}{\epsilon} \to \infty$.

(a) Obtain the first two terms of the outer expansion $\begin{pmatrix} X \\ Y \end{pmatrix}$.

(b) Obtain the system for $\begin{pmatrix} u \\ v \end{pmatrix}$.

(c) Determine the uniform approximation

$$x(t, \epsilon) = X_0(t) + O(\epsilon)$$

$$y(t, \epsilon) = Y_0(t) + v_0\left(\frac{t}{\epsilon}\right) + O(\epsilon).$$

5. Consider the initial value problem for the *conservation* equation

$$\epsilon \frac{d^2 x}{dt^2} = f(x)$$

with $x(0)$ and $\frac{dx}{dt}(0)$ prescribed. (An example is the pendulum equation $\epsilon \ddot{x} + \sin(\pi x) = 0$.) Consider an asymptotic solution

$$x(t, \epsilon) = X(t, \epsilon) + \epsilon u(\tau, \epsilon)$$

where $u \to 0$ as $\tau = \frac{t}{\epsilon} \to \infty$ and $0 \le t \le T < \infty$. Use Tikhonov–Levinson theory on the corresponding slow-fast system

$$\frac{dx}{dt} = y$$

$$\epsilon \frac{dy}{dt} = f(x)$$

under appropriate conditions.

(v) Remarks

In the remainder of this section, we will survey some important results from the literature. Readers should consult the references for further details.

We note that the typical requirements of the classical existence-uniqueness theory do not hold for the singular perturbation systems under consideration since their Lipschitz constant becomes unbounded when ϵ tends to zero. Sophisticated estimates are, nonetheless, provided by Nipp and Stoffer [351]. Needed asymptotic techniques, presented in Wasow [513], are updated in Hsieh and Sibuya [220] through the introduction of Gevrey asymptotics (cf. Ramis [405]) and Balser [25]). A formal power series $\sum_{m=0}^{\infty} a_m x^m$ is defined to be of *Gevrey order* s if there exist nonnegative numbers C and A such that

$$|a_m| \le C(m!)^s A^m$$

for all m (cf. Sibuya [457], Sibuya [458], and Canalis-Durand et al. [68]).

Fruchard and Schäfke [165] develop a "composite asymptotic expansion" approach which justifies matched asymptotic expansions for a class of ordinary differential equations, allowing some turning points. Their outer solutions and initial layer corrections are obtained as Gevrey expansions.

Instead of assuming asymptotic stability of the limiting fast system (the preceding hypothesis (ii)), one might instead consider the possibility of having rapid oscillations for the solution of the fast system (cf. Artstein et al. [13, 14]). It is useful, indeed, to interpret these solutions in terms of *Young measures*.

In his study of the *quasi-static state* analysis, Hoppensteadt [213, 214] considers the perturbed gradient system

$$\begin{cases} \frac{dx}{dt} & = f(x, y, t, \epsilon) \\ \epsilon\frac{dy}{dt} & = -\nabla_y\, G(x, y) + \epsilon g(x, y, t, \epsilon). \end{cases} \tag{3.98}$$

Since

$$\frac{dG}{dt} = -\frac{1}{\epsilon}\,\nabla_y\, G \cdot \nabla_y G + \nabla_x G \cdot f + \nabla_y G \cdot g = -\frac{1}{\epsilon}|\nabla_y\, G|^2 + O(1),$$

we might expect (under natural assumptions) the fast vector y to tend rapidly to an isolated minimum y^* of the energy $G(x, y)$, presuming $y(0)$ is in its domain of attraction. The corresponding limiting slow variable will satisfy

$$\frac{dx}{dt} = f(x, y^*, t, 0). \tag{3.99}$$

Extensions to more complicated systems are also given, including a four-dimensional *Lorenz* model

$$\begin{cases} \dot{x}_1 & = x_2 x_3 - b x_1 + \epsilon f_1(x, y, t, \epsilon) \\ \dot{x}_2 & = -x_1 x_3 + r x_3 - x_2 + \epsilon f_2(x, y, t, \epsilon) \\ \dot{x}_3 & = \sigma(x_2 - x_3) + \epsilon f_3(x, y, t, \epsilon) \\ \epsilon\dot{y} & = \lambda y - y^3 + \epsilon g(x, y, t, \epsilon) \end{cases}$$

that has the function

$$W(y) = \frac{1}{2}(y - \sqrt{\lambda})^2$$

as a *Liapunov* or *energy function* for the branch $y = \sqrt{\lambda} + O(\epsilon)$ (with $\lambda > 0$) of the limiting fast system (cf. Brauer and Nohel [59]). Solutions beginning nearby remain close to the manifold, and may exhibit chaotic behavior for certain values of the parameters b, r, and σ.

Asymptotic expansions, as in the ansatz (3.70), are used in Hairer and Wanner [192] to develop *Runge–Kutta* methods for numerically integrating vector initial value problems in the singularly perturbed form

$$\begin{cases} \dot{x} = f(x, y), \\ \epsilon\dot{y} = g(x, y), \end{cases} \tag{3.100}$$

assuming that the Jacobian matrix g_y is stable near the solution of the 'reduced differential-algebraic system. Note, in the planar situation, that trajectories will satisfy

$$\epsilon \frac{dy}{dx} = \frac{g(x,y)}{f(x,y)}.$$

In particular, Hairer and Wanner begin their treatment of such *stiff* differential equations by considering the one-dimensional example

$$\dot{y} = g(x,y) = -50(y - \cos x)$$

from Curtiss and Hirschfelder [107] (with $\epsilon = 0.02$), pointing out the spurious oscillations one finds with the explicit Euler method, in contrast to the success obtained using the *backward differentiation* formula

$$y_{n+1} - y_n = hg(x_{n+1}, y_{n+1}). \tag{3.101}$$

Aiken [5] provides a review of the early literature from the chemical engineering perspective.

The existence of periodic solutions to the slow-fast vector system

$$\begin{cases} \dot{x} = f(x,y,t,\epsilon) \\ \epsilon\dot{y} = g(x,y,t,\epsilon) \end{cases} \tag{3.102}$$

was considered by Flatto and Levinson [149] and generalized in Wasow [513]. We will assume that f and g are periodic in t with period ω and that the reduced system

$$\begin{cases} \dot{X}_0 = f(X_0, Y_0, t, 0) \\ 0 = g(X_0, Y_0, t, 0) \end{cases} \tag{3.103}$$

has a solution $\begin{pmatrix} X_0 \\ Y_0 \end{pmatrix}$ of period ω. The question is whether or not the full system (3.102) has a nearby periodic solution of the same period.

Let s be the vector parameter of initial values for X_0, with the corresponding variational system

$$\begin{cases} \frac{d}{dt}\left(\frac{\partial X_0}{\partial s}\right) = f_x(X_0, Y_0, t, 0)\frac{\partial X_0}{\partial s} + f_y(X_0, Y_0, t, 0)\frac{\partial Y_0}{\partial s} \\ 0 = g_x(X_0, Y_0, t, 0)\frac{\partial X_0}{\partial s} + g_y(X_0, Y_0, t, 0)\frac{\partial Y_0}{\partial s} \end{cases} \tag{3.104}$$

We will assume that

(i) there is a smooth nonsingular matrix $P(t)$ of period ω so that

$$P^{-1}(t)g_y(X_0, Y_0, t, 0)P(t) \equiv \begin{pmatrix} B(t) & 0 \\ 0 & -C(t) \end{pmatrix} \tag{3.105}$$

with $B(t)$ and $C(t)$ being stable matrices.

Since $g_y(X_0, Y_0, t, 0)$ is nonsingular and

$$\frac{\partial Y_0}{\partial s} = -g_y^{-1}(X_0, Y_0, t, 0)g_x(X_0, Y_0, t, 0)\frac{\partial X_0}{\partial s}, \qquad (3.106)$$

$\xi \equiv \frac{\partial X_0}{\partial s}$ will satisfy the linear system

$$\frac{d\xi}{dt} = A(t)\xi \qquad (3.107)$$

for $A(t) \equiv f_x(X_0, Y_0, t, 0) - f_y(X_0, Y_0, t, 0)g_y^{-1}(X_0, Y_0, t, 0)g_x(X_0, Y_0, t, 0)$.
We will also assume

(ii) the variational equation (3.107) has no nontrivial solution of period ω.

(Recall *Floquet theory* and the *Fredholm alternative theorem* from Codding-ton and Levinson [91]). Flatto and Levinson [149] show that the full system (3.102) will then have a solution of period ω with a uniform asymptotic expansion

$$\begin{pmatrix} x(t, \epsilon) \\ y(t, \epsilon) \end{pmatrix} \sim \sum_{k=0}^{\infty} \begin{pmatrix} X_k(t) \\ Y_k(t) \end{pmatrix} \epsilon^k. \qquad (3.108)$$

Because there are no distinguished boundary points, the periodic solution doesn't need boundary layers.

Verhulst [505] considers the scalar Riccati example

$$\epsilon\dot{y} = a(t)y - y^2 \qquad (3.109)$$

with $a(t)$ positive and periodic. The reduced problem has a nontrivial and stable periodic solution

$$Y_0(t) = a(t)$$

while we suppose the singularly perturbed equation (3.109) has a regularly perturbed solution

$$y(t, \epsilon) = Y_0(t) + \epsilon Y_1(t) + \dots$$

This requires

$$\dot{Y}_0 = a(t)Y_1 - 2Y_0Y_1$$

at $O(\epsilon)$ order, so $Y_1 = -\frac{\dot{a}(t)}{a(t)}$ implies the corresponding periodic approximation

$$y(t, \epsilon) \sim a(t) - \epsilon\frac{\dot{a}(t)}{a(t)} + \dots. \qquad (3.110)$$

Kopell and Howard [258] studied the *Belousov–Zhabotinsky reaction*, which provides dramatic chemical oscillations with color changes. When one seeks a traveling wave solution, a concentration C satisfies a singularly perturbed differential equation

$$C' = F(C) + \beta C''$$

with a small $\beta > 0$. Kopell [257] supposes that the reduced problem has a stable limit cycle and she seeks a nearby invariant manifold for the perturbed problem. This provides a major motivation for Fenichel's geometric theory from 1979, which generalizes Anosov [11].

The concept of a slow integral manifold (cf. Wiggins [522], Nipp and Stoffer [351], Goussis [178], Shchepakina et al. [450], Kuehn [268], and Roberts [418]) is valuable in many applied contexts, including chemical kinetics, control theory, and computation (cf. also, Kokotović et al. [256] and Gear et al. [167]). Let's again consider the initial value problem for the slow-fast $m + n$ dimensional system

$$\begin{cases} \dot{x} = f(x, y, t, \epsilon) \\ \epsilon \dot{y} = g(x, y, t, \epsilon) \end{cases} \tag{3.111}$$

on $t \geq 0$, subject to the usual Tikhonov–Levinson stability hypotheses. We will determine a corresponding *slow manifold* described by

$$y(t, \epsilon) = h(x(t, \epsilon), t, \epsilon) \tag{3.112}$$

for a vector function h to be determined termwise as a power series in ϵ. Motion along it will then be governed by the m-dimensional slow system

$$\dot{x} = f(x, h(x, t, \epsilon), t, \epsilon), \tag{3.113}$$

subject to the prescribed initial vector $x(0)$. This approach provides a substantial reduction in dimensionality when n is large, although it fails to describe the usual rapid nonuniform convergence of y in the $O(\epsilon)$-thick initial layer. However, in chemical kinetics, for example, the initial layer behavior may occur too quickly to measure in the lab. Thus, it's then natural to seek such a quasi-steady state. Still, the fast equation and the chain rule applied to (3.112) imply the *invariance equation*

$$\epsilon \dot{y} = \epsilon \left(\frac{\partial h}{\partial x} f + \frac{\partial h}{\partial t} \right) = g. \tag{3.114}$$

To lowest order, this requires

$$g(x, h_0, t, 0) = 0, \tag{3.115}$$

so we naturally take

$$h_0 = \phi(x, t) \tag{3.116}$$

to be an isolated root of the limiting fast system (3.115). Moreover, we again require the root ϕ to be attractive in the sense that

$$g_y(x, \phi(x, t), t, 0)$$

is a *strictly stable* matrix, thereby ruling out any repulsive roots that might occur. Higher-order terms in the expansion

$$h(x, t, \epsilon) = \phi(x, t) + \epsilon h_1(x, t) + \dots \tag{3.117}$$

follow readily since $g(x, \phi(x,t), t, 0) = 0$ implies the expansion $g(x, h(x, t, \epsilon), t, \epsilon) = g_y(x, \phi(x, t), t, 0)(\epsilon h_1(x, t) + \ldots) + \epsilon g_\epsilon(x, \phi(x, t), t, 0) + \ldots = 0$ about $\epsilon = 0$. Balancing the $O(\epsilon)$ terms in (3.114) then implies that

$$\frac{\partial h}{\partial x}(x, \phi(x, t), t, 0) f(x, \phi(x, t), t, 0) + \frac{\partial h}{\partial t}(x, \phi(x, t), t, 0)$$

$$= g_y(x, \phi(x, t), t, 0) h_1(x, t) + g_\epsilon(x, \phi(x, t), t, 0). \tag{3.118}$$

This specifies h_1 since g_y is nonsingular and all else is known. h_2 next follows analogously from the $O(\epsilon^2)$ terms in (3.114). Thus, it is convenient to describe the slow manifold in terms of the outer limit, avoiding the initial layer correction.

Examples

1. Kokotović et al. [256] considered an initial value problem like

$$\begin{cases} \dot{x} = f(x, y, t) \equiv txy, & x(0) = 1 \\ \epsilon \dot{y} = g(x, y, t) \equiv -(y - 4)(y - 2)(y + tx), & y(0) \text{ given.} \end{cases} \tag{3.119}$$

We naturally anticipate having an asymptotic solution of the form

$$\begin{cases} x(t, \epsilon) = X(t, \epsilon) + \epsilon u(\tau, \epsilon) \\ y(t, \epsilon) = Y(t, \epsilon) + v(\tau, \epsilon) \end{cases}$$

with an outer solution $\begin{pmatrix} X \\ Y \end{pmatrix}$ and an initial layer correction $\begin{pmatrix} \epsilon u \\ v \end{pmatrix}$ that tends to zero as $\tau = t/\epsilon$ tends to infinity. The outer limit $\begin{pmatrix} X_0 \\ Y_0 \end{pmatrix}$ will then satisfy the reduced problem

$$\begin{cases} \dot{X}_0 = tX_0Y_0, & X_0(0) = 1 \\ 0 = -(Y_0 - 4)(Y_0 - 2)(Y_0 + tX_0). \end{cases} \tag{3.120}$$

The first possibility

$$Y_0(t) = 4, \quad \dot{X}_0 = 4tX_0, \quad X_0(0) = 1$$

for the root Y_0 determines the bounded outer limit

$$\begin{pmatrix} X_0(t) \\ Y_0(t) \end{pmatrix} = \begin{pmatrix} e^{2t^2} \\ 4 \end{pmatrix} \tag{3.121}$$

for finite t. It provides the complete outer expansion and thereby the corresponding stable integral manifold. For $y(0) \neq 4$, however, we need a nontrivial boundary layer correction at $t = 0$. Its leading term v_0 must then satisfy

$$\frac{dv_0}{d\tau} = g(x(0), Y_0(0) + v_0, 0, 0) = -v_0(v_0 + 2)(v_0 + 4), \quad v_0(0) = y(0) - 4$$
(3.122)

and must decay to zero as $\tau \to \infty$. Checking the sign of $\frac{dv_0}{d\tau}$ shows that we will need $v_0(0) > -2$ or $y(0) > 2$ in order to attain such asymptotic stability.

The second possibility

$$Y_0(t) = 2$$
(3.123)

provides $X_0(t) = e^{t^2}$. For $y(0) \neq 2$, we will need a nontrivial limiting initial layer correction $v_0(\tau)$ satisfying

$$\frac{dv_0}{d\tau} = -(v_0 - 2)v_0(v_0 + 2), \quad v_0(0) = y(0) - 2.$$
(3.124)

Its trivial rest point is, however, unstable, as is the corresponding integral manifold. Thus, we rule out (3.123), except when $y(0) = 2$ exactly.

Finally, when we take

$$Y_0(t) = -tX_0(t),$$
(3.125)

$\dot{X}_0 = tX_0Y_0 = -t^2 X_0^2$, and $X_0(0) = 1$ determine the limiting outer solution

$$\begin{pmatrix} X_0(t) \\ Y_0(t) \end{pmatrix} = \begin{pmatrix} \frac{3}{t^2+3} \\ \frac{-3t}{t^2+3} \end{pmatrix}.$$
(3.126)

The next term in the outer expansion must then satisfy the linear system

$$\begin{cases} \dot{X}_1 = t(X_1Y_0 + X_0Y_1) = -t^2 X_0 X_1 + t X_0 Y_1 \\ \dot{Y}_0 = -[Y_1(Y_0 - 2) + (Y_0 - 4)Y_1](Y_0 + tX_0) - (Y_0 - 4)(Y_0 - 2)(Y_1 + tX_1) \end{cases}$$

so

$$Y_1 + tX_1 = \frac{-\dot{Y}_0}{Y_0^2 - 6Y_0 + 8} = \frac{X_0 - t^3 X_0^2}{t^2 X_0^2 + 6t X_0 + 8}$$
(3.127)

where

$$\dot{X}_1 = -2t^2 X_0 X_1 + \frac{t X_0^2(1 - t^3 X_0)}{t^2 X_0^2 + 6t X_0 + 8}, \quad X_1(0) = u_0(0).$$
(3.128)

The corresponding limiting initial layer system

$$\begin{cases} \frac{du_0}{d\tau} = 0 \\ \frac{dv_0}{d\tau} = -(v_0 - 4)(v_0 - 2)v_0, \quad v_0(0) = y(0) \end{cases}$$
(3.129)

has the trivial rest point provided $y(0) < 2$. In summary, we obtain one of the possible asymptotic solutions depending on the sign of $y(0) - 2$. The solution lies on an attractive slow invariant manifold when $y(0) = 4$ or 0.

The *geometric* singular perturbation theory of Fenichel [146] generalizes Tikhonov–Levinson theory by replacing its first stability assumption by *normal hyperbolicity*. See Fenichel [147], Kaper [233], Jones and Khibnik [227], Krupa and Szmolyan [265], Verhulst and Bakri [505], Kosiak and Szmolyan [260], and Kuehn [268] for updated treatments. In particular, then, imaginary eigenvalues of g_y are not allowed, but unstable eigenvalues are. (Interestingly, Neil Fenichel's work was "ahead of its time." It didn't attract the attention it merited for many years.) Hastings and McLeod [198] include several applications of Fenichel's theory, which they simplify. A large variety of sophisticated approaches are combined in Desroches et al. [116]. Other significant extensions of Tikhonov's theorem include Nipp [349] (cf., also, Nipp and Stoffer [351]).

The situation where the first Tikhonov–Levinson stability assumption is violated because the Jacobian matrix g_y is everywhere singular might be called a *singular singular perturbation* problem (cf. Gu et al. [187]). Narang-Siddarth and Valasek [340] say these are in nonstandard form.

2. A simple example of a singular problem is provided by the linear initial value problem

$$\epsilon \dot{y} = A(\epsilon)y, \quad y(0) = \begin{pmatrix} 1 \\ 1 \end{pmatrix} \tag{3.130}$$

for the nearly singular state matrix

$$A(\epsilon) = \begin{pmatrix} 1 - 2\epsilon & 2 - 2\epsilon \\ -1 + \epsilon & -2 + \epsilon \end{pmatrix} \tag{3.131}$$

with eigenvalues -1 and $-\epsilon$ and corresponding eigenvectors $\begin{pmatrix} 1 \\ -1 \end{pmatrix}$ and $\begin{pmatrix} 2 \\ -1 \end{pmatrix}$. Applying the initial condition provides the exact solution

$$y(t, \epsilon) = \begin{pmatrix} 4 \\ -2 \end{pmatrix} e^{-t} + \begin{pmatrix} -3 \\ 3 \end{pmatrix} e^{-t/\epsilon} \quad \text{for} \quad t \geq 0 \tag{3.132}$$

in the anticipated form

$$y(t, \epsilon) \sim Y_0(t) + \xi_0(\tau)$$

for an outer solution $Y_0(t)$ and an initial layer correction $\xi_0(\tau)$ that decays to zero as $\tau = t/\epsilon \to \infty$.

If we, instead, simply sought an outer solution

$$Y(t, \epsilon) \sim \sum_{j \geq 0} Y_j(t) \epsilon^j \tag{3.133}$$

of $\epsilon \dot{y} = A(t)y$ with $Y_j = \begin{pmatrix} Y_{1j} \\ Y_{2j} \end{pmatrix}$, the leading terms require that

$$Y_{10} + 2Y_{20} = 0, \tag{3.134}$$

but leaves Y_0 otherwise unspecified. At $O(\epsilon)$, we'd need

$$\begin{cases} \dot{Y}_{10} = Y_{11} + 2Y_{21} - 2Y_{10} - 2Y_{20} \\ \text{and} \\ \dot{Y}_{20} = -Y_{11} - 2Y_{21} + Y_{10} + Y_{20}. \end{cases} \tag{3.135}$$

Adding implies that

$$\dot{Y}_{10} + \dot{Y}_{20} = -Y_{10} - Y_{20}.$$

Because $Y_{10} = -2Y_{20}$, however, $\dot{Y}_{20} = -Y_{20}$ so

$$Y_0(t) = \begin{pmatrix} -2 \\ 1 \end{pmatrix} e^{-t} k_0 \tag{3.136}$$

for a constant k_0 to be determined by matching.

More directly, we could change variables by putting $A(\epsilon)$ in a more convenient triangular form. Let us set

$$z = \begin{pmatrix} z_1 \\ z_2 \end{pmatrix} \equiv Py = \begin{pmatrix} 1 & -1 \\ 1 & 1 \end{pmatrix} y, \tag{3.137}$$

so

$$y = \frac{1}{2} \begin{pmatrix} 1 & 1 \\ -1 & 1 \end{pmatrix} z$$

and the initial value problem (3.130) is transformed to

$$\begin{cases} \epsilon \dot{z}_1 = -z_1 + 3(1 - \epsilon)z_2, & z_1(0) = 0 \\ \dot{z}_2 = -z_2, & z_2(0) = -2, \end{cases} \tag{3.138}$$

a problem in fast-slow form that can be uniquely solved using Tikhonov–Levinson theory. We get

$$\begin{cases} z_1(t) = 6e^{-t} - 6e^{-t/\epsilon} \\ \text{and} \\ z_2(t) = 2e^{-t}, \end{cases} \tag{3.139}$$

corresponding to the constant $k_0 = 3$ in (3.136). Although only the first component of z has an initial layer, both components of y do.

3. A nonlinear example is given by

$$\begin{cases} \epsilon \dot{y}_1 = y_1 + y_2 - \frac{1}{2}(y_1 + y_2)^3 + \frac{\epsilon}{\sqrt{2}}(y_1^2 - y_2^2), & y_1(0) = 2 \\ \epsilon \dot{y}_2 = y_1 + y_2 - \frac{1}{2}(y_1 + y_2)^3 - \frac{\epsilon}{\sqrt{2}}(y_1^2 - y_2^2), & y_2(0) = -2. \end{cases} \tag{3.140}$$

Now, the reduced problem

$$Y_{10} + Y_{20} - \frac{1}{2}(Y_{10} + Y_{20})^3 = 0 \tag{3.141}$$

has the three families of solutions

$$Y_{10} + Y_{20} = 0 \quad \text{or} \quad \pm\sqrt{2}. \tag{3.142}$$

Tikhonov–Levinson theory doesn't apply, but if we transform the problem by setting

$$z = \begin{pmatrix} 1 & 1 \\ 1 & -1 \end{pmatrix} y, \tag{3.143}$$

we get the separated fast-slow system

$$\begin{cases} \epsilon \dot{z}_1 = z_1 - \frac{1}{2} z_1^3, & z_1(0) = -2 \\ \dot{z}_2 = \sqrt{2} z_1 z_2, & z_2(0) = 0. \end{cases} \tag{3.144}$$

We immediately integrate the Bernoulli equation for z_1 to get

$$z_1(t,\epsilon) = -\sqrt{\frac{2}{1 - \frac{1}{2} e^{-2t/\epsilon}}} \tag{3.145}$$

and we conveniently rewrite it in the anticipated form

$$z_1(t,\epsilon) = -\sqrt{2} + u_1(\tau) \tag{3.146}$$

with the outer solution $-\sqrt{2}$ and the decaying initial layer correction

$$u_1(\tau) = \sqrt{2}\left(1 - \frac{1}{\sqrt{1 - \frac{1}{2} e^{-2\tau}}}\right). \tag{3.147}$$

Integrating the remaining linear equation for z_2, we get

$$\begin{aligned} z_2(t,\tau,\epsilon) &= -2e^{-2t} e^{\sqrt{2}\epsilon \int_0^\tau u_1(r)dr} \\ &\equiv Z_2(t,\epsilon) e^{-\sqrt{2}\epsilon \int_\tau^\infty u_1(r)dr}. \end{aligned} \tag{3.148}$$

Setting

$$z_2(t,\epsilon) = Z_2(t,\epsilon) + \epsilon u_2(\tau,\epsilon), \tag{3.149}$$

we have a decaying initial layer correction $\epsilon u_2(\tau,\epsilon)$. Power series for Z_2 and u_2 can be obtained termwise. The limiting outer solution

$$Z_1(t,0) = -\sqrt{2} \quad \text{and} \quad Z_2(t,0) = -2e^{-2t}$$

corresponds to the outer limits

$$\begin{cases} Y_{10}(t) = -\frac{1}{\sqrt{2}} - e^{-2t} \\ \text{and} \\ Y_{20}(t) = -\frac{1}{\sqrt{2}} + e^{-2t} \end{cases} \tag{3.150}$$

and to the conserved constant

$$Y_{10}(t) + Y_{20}(t) = -\sqrt{2}. \tag{3.151}$$

Since neither $Y_{10}(0) = -2$ nor $Y_{20}(0) = -2$, both components of y need initial layer corrections.

For the nonlinear n-dimensional initial value problem

$$\epsilon \dot{y} = g(y, t, \epsilon), \quad t \geq 0, \tag{3.152}$$

we expect the limiting solution to satisfy

$$g(Y_0, t, 0) = 0.$$

When g_y is singular with a constant rank $0 < k < n$ and when its nontrivial eigenvalues are stable, we might seek additional constraints on the outer limit Y_0 by differentiating $g = 0$. (Recall the related concept of the *index* of a differential-algebraic equation (cf. Ascher and Petzold [15] and Lamour et al. [279]). Shchepakina et al. [450] describe applications, including singular ones and bimolecular reactions.

Historical Remark

Tikhonov made important contributions to many fields of mathematics, including topology and cybernetics. He also rose to the top of the Communist Party hierarchy in the Soviet Union, attaining great power and exerting his anti-Semitism (like Pontryagin) by, for example, influencing the results of entrance exams at Moscow State University.

Levinson, as the child of poor Russian Jewish immigrants in Revere, Massachusetts, naturally supported leftish causes. Norbert Wiener recognized his brilliance and got him (with some help from Hardy) a faculty position at the Massachusetts Institute of Technology. (Harvard was, presumably, unwilling to hire Jewish mathematicians in 1937.) In the McCarthy era of Communist witchhunts, Levinson was called to Washington to testify, but he refused to "name names" (cf. Levinson [288] and O'Connor and Robertson [355]).

(c)   Two-Point Problems

The linear first-order scalar equation

$$\epsilon y' + a(x)y = b(x), \quad x \geq 0 \tag{3.153}$$

has the exact solution

$$y(x, \epsilon) = e^{-\frac{1}{\epsilon} \int_0^x a(s)\, ds}\, y(0) + \frac{1}{\epsilon} \int_0^x e^{-\frac{1}{\epsilon} \int_s^x a(t)\, dt}\, b(s)\, ds.$$

For bounded x and smooth coefficients, we can use repeated integrations by parts when

$$a(x) > 0$$

to show that y has a generalized asymptotic expansion of the form

$$y(x, \epsilon) \sim A(x, \epsilon) + B(\epsilon)e^{-\frac{1}{\epsilon}\int_0^x a(s)\, ds} \tag{3.154}$$

for power series A and B. For example, since

$$\frac{1}{\epsilon}\int_0^x e^{-\frac{1}{\epsilon}\int_s^x a(t)\, dt}b(s)\, ds \sim \frac{b(x)}{a(x)} - e^{-\frac{1}{\epsilon}\int_0^x a(t)\, dt}\frac{b(0)}{a(0)},$$

$A(x, 0) = \frac{b(x)}{a(x)}$ and $B(0) = y(0) - \frac{b(0)}{a(0)}$. Indeed, the series can be found directly by regular perturbation methods (using undetermined coefficients in the power series for A and B). Since there is an initial layer near $x = 0$, we could also introduce the stretched variable

$$\xi = x/\epsilon$$

and expand the product $B(\epsilon)e^{-\frac{1}{\epsilon}\int_0^{\epsilon\xi} a(s)\, ds}$ in its Maclaurin expansion about $\epsilon = 0$ to find the composite asymptotic solution

$$y(x, \epsilon) = A(x, \epsilon) + C(\xi, \epsilon) \tag{3.155}$$

for the same outer solution $A(x, \epsilon)$, where the coefficients of the initial layer correction

$$C(\xi, \epsilon) \sim \sum_{k \geq 0} C_k(\xi)\epsilon^k$$

tend to zero as $\xi \to \infty$. Clearly, the expansion (3.154) is preferable, because it provides more immediate details regarding boundary layer behavior. In particular, it explicitly shows that the nonuniform behavior in the initial layer depends on the stretched variable

$$\eta = \frac{1}{\epsilon}\int_0^x a(s)\, ds,$$

rather than its local limit $a(0)\xi$. Indeed, $e^{-\eta}$ exactly satisfies the homogeneous equation. As we will later find, the expansion (3.155) corresponds to matching and (3.154) to two-timing.

For the linear second-order equation

$$\epsilon y'' + a(x)y' + b(x)y = c(x) \tag{3.156}$$

with $a(x) > 0$, we cannot generally write down the exact solution (unless we happen to know a nontrivial solution of the homogeneous equation). Nonetheless, we will find that the asymptotic solution of the two-point problem with

$$y(0) \quad \text{and} \quad y(1) \quad \text{prescribed}$$

will likewise have the asymptotic form

$$y(x, \epsilon) \sim A(x, \epsilon) + B(x, \epsilon) e^{-\frac{1}{\epsilon} \int_0^x a(s) \, ds} \tag{3.157}$$

(corresponding to WKB theory) where the outer expansion $A(x, \epsilon)$ will now be a regular power series solution of the terminal value problem

$$\epsilon A'' + a(x) A' + b(x) A = c(x), \quad A(1, \epsilon) = y(1) \tag{3.158}$$

and where $B(x, \epsilon)$ will be a regular power series solution of the initial value problem

$$\epsilon B'' - aB' - a'B + bB = 0, \quad B(0, \epsilon) = y(0) - A(0, \epsilon) \tag{3.159}$$

(since the product $Be^{-\frac{1}{\epsilon} \int_0^x a(s) \, ds}$ must satisfy the homogeneous differential equation). Curiously, the differential equation for B is the *adjoint* of that for A (when c is zero).

Analogous (though somewhat more complicated) results hold for the nonlinear equation

$$\epsilon y'' + a(x) y' + f(x, y) = 0 \tag{3.160}$$

again for Dirichlet boundary conditions. Before obtaining them, let us first show how the more familiar method of *boundary layer corrections* works. With

$$a(x) > 0 \quad \text{on} \quad 0 \leq x \leq 1,$$

we would naturally expect an initial layer of nonuniform convergence when $y(0)$ and $y(1)$ are prescribed. Thus, for smooth coefficients a and f, we will seek a composite asymptotic expansion of the form

$$y(x, \epsilon) = Y(x, \epsilon) + v(\xi, \epsilon) \tag{3.161}$$

where $v \to 0$ as the stretched coordinate

$$\xi = \frac{x}{\epsilon}$$

$\to \infty$ and where the outer expansion Y and the initial layer correction v have power series expansions

$$Y(x, \epsilon) \sim \sum_{k \geq 0} Y_k(x) \epsilon^k \quad \text{and} \quad v(\xi, \epsilon) \sim \sum_{k \geq 0} v_k(\xi) \epsilon^k.$$

Away from $x = 0$, $y \sim Y$ to all orders (and, likewise, for its derivatives), so Y must satisfy

$$\epsilon Y'' + a(x) Y' + f(x, Y) = 0, \quad Y(1, \epsilon) = y(1). \tag{3.162}$$

Clearly, Y_0 must satisfy the nonlinear reduced problem

$$a(x) Y_0' + f(x, Y_0) = 0, \quad Y_0(1) = y(1). \tag{3.163}$$

Assuming that its solution Y_0 exists from $x = 1$, back to $x = 0$, later Y_ks must satisfy linearized problems

$$a(x)Y_k'' + f_x(x, Y_0)Y_k + \alpha_{k-1}(x) = 0, \quad Y_k(1) = 0 \qquad (3.164)$$

there, where each α_{k-1} is known successively in terms of earlier coefficients Y_j and their first two derivatives. Using an integrating factor, each Y_k then follows uniquely throughout the interval. It would be unlikely that $Y(0, \epsilon) = y(0)$, however, so a nontrivial corrector v must be expected.

Knowing Y asymptotically, $y' = Y' + \frac{1}{\epsilon}\frac{dv}{d\xi}$ and $\epsilon y'' = \epsilon Y'' + \frac{1}{\epsilon}\frac{d^2v}{d\xi^2}$ imply that the initial layer correction v must satisfy the differential equation

$$\frac{d^2v}{d\xi^2} + a(\epsilon\xi)\frac{dv}{d\xi} + \epsilon\Big(f\big(\epsilon\xi, Y(\epsilon\xi, \epsilon) + v(\xi, \epsilon)\big) - f\big(\epsilon\xi, Y(\epsilon\xi, \epsilon)\big)\Big) = 0, \quad (3.165)$$

the initial condition

$$v(0, \epsilon) = y(0) - Y(0, \epsilon), \qquad (3.166)$$

and decay to zero as $\xi \to \infty$. Thus, the leading coefficient v_0 must satisfy the linear problem

$$\frac{d^2v_0}{d\xi^2} + a(0)\frac{dv_0}{d\xi} = 0, \quad v_0(0) = y(0) - Y_0(0) \quad \text{and} \quad v_0 \to 0 \text{ as } \xi \to \infty,$$

so

$$v_0(\xi) = e^{-a(0)\xi}(y(0) - Y_0(0)). \qquad (3.167)$$

Next, we will need

$$\frac{d^2v_1}{d\xi^2} + a(0)\frac{dv_1}{d\xi} + a'(0)\xi\frac{dv_0}{d\xi} + f(0, Y_0(0) + v_0(\xi)) - f(0, Y_0(0)) = 0,$$

$$v_1(0) = -Y_1(0) \quad \text{and} \quad v_1 \to 0 \quad \text{as} \quad \xi \to \infty.$$

The unique solution

$$v_1(\xi) = -e^{-a(0)\xi}Y_1(0) - \int_0^\xi e^{a(0)(s-\xi)}\big[f(0, Y_0(0) + v_0(\xi)) \\ - f(0, Y_0(0)) - a'(0)a(0)\, s\, v_0(s)\big]\, ds \qquad (3.168)$$

decays like $\xi e^{-a(0)\xi}$ as $\xi \to \infty$. Subsequent v_js follow analogously, in turn. We will later obtain a somewhat more satisfying solution using multiscale methods with slow and fast variables x and $\eta = \frac{1}{\epsilon}\int_0^x a(s)\,ds$. Numerical methods for such problems are presented in Roos et al. [419] and Ascher et al. [16]. Related techniques for partial differential equations are given in Shishkin and Shishkina [452], Linss [294], and Miller et al. [317]. The variety of two-point singular perturbation problems one can confidently solve numerically is, sadly, quite limited, compared to the success found for stiff initial value problems. This appropriately remains a topic of substantial current research and importance. Useful recommendations about software for solving singularly perturbed two-point problems can be found on the home-page of Professor Jeff Cash of Imperial College, London (cf. also, Soetaert et al. [469]).

If we consider the two-point problem for the scalar Liénard equation

$$\epsilon y'' + f(y)y' + g(y) = 0, \qquad 0 \leq x \leq 1 \qquad (3.169)$$

with $y(0)$ and $y(1)$ prescribed, we can again expect to have an asymptotic solution of the form

$$y(x, \epsilon) = Y(x, \epsilon) + v(\xi, \epsilon) \qquad (3.170)$$

provided

(i) the reduced problem

$$f(Y_0)Y_0' + g(Y_0) = 0, \qquad Y_0(1) = y(1) \qquad (3.171)$$

has a solution $Y_0(x)$ on $0 \leq x \leq 1$ with

$$f(Y_0) > 0.$$

(Note the monotonic implicit solution $x - 1 = \int_{Y_0(x)}^{y(1)} \frac{f(r)}{g(r)} \, dr$.)

and

(ii) the linear integrated initial layer problem

$$\frac{dv_0}{d\xi} + f(Y_0(0))v = 0, \qquad v_0(0) = y(0) - Y_0(0) \qquad (3.172)$$

has a solution $v_0(\xi)$ on $\xi \geq 0$ that decays to zero as $\xi \equiv \frac{x}{\epsilon} \to \infty$. (This simply requires $f(Y_0(0)) > 0$ since the solution is an exponential.)

Treating the problem with $f(Y_0) < 0$ proceeds analogously, using a terminal layer, but real complications arise when $f(Y_0)$ has a zero within the interval.

As a specific example, suppose

$$\epsilon y'' = 2yy' \qquad \text{with} \qquad y(1) < 0 \text{ and } y(0) + y(1) < 0. \qquad (3.173)$$

Then, we obtain the attractive constant outer solution

$$Y(x, \epsilon) = y(1) < 0 \qquad (3.174)$$

while the supplementary initial layer correction $v(\xi, \epsilon)$ must be a decaying solution of $v_{\xi\xi} = 2(y(1) + v)v_\xi$. Integrating from infinity, we must satisfy the Riccati equation

$$v_\xi - 2y(1)v - v^2 = 0, \qquad v(0) = y(0) - y(1). \qquad (3.175)$$

With the assumed sign restrictions, v exists and decays to zero as $\xi \to \infty$.

Cole [92] considered the linear problem

$$\epsilon y'' + \sqrt{x}\,xy' - y = 0, \quad y(0) = 0 \quad \text{and} \quad y(1) = e^2, \tag{3.176}$$

with an initial turning point. Because $\sqrt{x} > 0$ for $x > 0$, we might stubbornly still seek an asymptotic solution of the composite form

$$y(x, \epsilon) = Y(x, \epsilon) + v(\xi, \epsilon^\beta), \tag{3.177}$$

with an outer expansion Y that satisfies the terminal value problem

$$\epsilon Y'' + \sqrt{x}\,Y' - Y = 0, \quad Y(1, \epsilon) = e^2 \tag{3.178}$$

as a power series in ϵ, and with an initial layer correction v satisfying the stretched equation

$$\epsilon^{1-2\alpha}\frac{d^2 v}{d\xi^2} + \epsilon^{-\alpha/2}\sqrt{\xi}\frac{dv}{d\xi} - v = 0, \tag{3.179}$$

the initial condition

$$v(0, \epsilon^\beta) = y(0) - Y(0, \epsilon), \tag{3.180}$$

and which decays to zero as the appropriate stretched variable

$$\xi = \frac{x}{\epsilon^\alpha}, \tag{3.181}$$

for some $\alpha > 0$, tends to infinity. We will take v to have a power series in ϵ^β, for a power $\beta > 0$ to be determined. The purpose of the new stretching ξ is to balance different terms in the differential equation (3.176) within the initial layer. The *dominant balance* argument (cf. Bender and Orszag [36] and Nipp [350]) here requires us to select α so that

$$1 - 2\alpha = -\frac{\alpha}{2} \quad \text{or} \quad \alpha = 2/3. \tag{3.182}$$

Since this leaves

$$\frac{d^2 v}{d\xi^2} + \sqrt{\xi}\frac{dv}{d\xi} - \epsilon^{1/3}v = 0, \tag{3.183}$$

we naturally take $\beta = 1/3$.

The outer expansion $Y \sim \sum_{k \geq 0} Y_k \epsilon^k$ for (3.176) must satisfy

$$\sqrt{x}\,Y_0' - Y_0 = 0, \quad Y_0(1) = e^2$$

and $\sqrt{x}\,Y_1' - Y_1 + Y_0'' = 0$, $Y_1(1) = 0$, so we get

$$Y(x, \epsilon) = e^{2\sqrt{x}}\left(1 + \epsilon\left(-\frac{1}{2x} + \frac{2}{\sqrt{x}} - \frac{3}{2}\right) + \dots\right). \tag{3.184}$$

Indeed, following a suggestion of E. Kirkinis, we can peel off $e^{2\sqrt{x}}$ by setting

$$y = e^{2\sqrt{x}} z. \tag{3.185}$$

The corresponding outer expansion $Z(x, \epsilon)$ in inner variables provides

$$Z(\epsilon^{2/3}\xi, \epsilon) = 1 - \frac{\epsilon^{1/3}}{2\xi} + \frac{2\epsilon^{2/3}}{\sqrt{\xi}} - \frac{3c}{2} + \ldots, \tag{3.186}$$

conveniently a power series in $\epsilon^{1/3}$. The transformed equation is

$$\epsilon z'' + \left(\sqrt{x} + \frac{2\epsilon}{\sqrt{x}} \right) z' + \epsilon \left(\frac{1}{x} - \frac{1}{2} \frac{1}{(\sqrt{x})^3} \right) z = 0$$

and the stretched equation for the corresponding inner solution $w(\xi, \epsilon^{1/3})$ in terms of $\xi = x/\epsilon^{2/3}$ is

$$\frac{d^2 w}{d\xi^2} + \left(\sqrt{\xi} + \frac{2\epsilon^{1/3}}{\sqrt{\xi}} \right) \frac{dw}{d\xi} + \left(-\frac{\epsilon^{1/3}}{2\xi^{3/2}} + \frac{\epsilon^{2/3}}{\xi} \right) w = 0, \quad w(0) = 0. \tag{3.187}$$

Expanding

$$w(\xi, \epsilon^{1/3}) \sim \sum_{k=0}^{\infty} w_k(\xi) \epsilon^{k/3} \tag{3.188}$$

and integrating the resulting initial value problems, we first obtain

$$w_0(\xi) = c_0 \int_0^{\xi} e^{-\frac{2}{3} s^{3/2}} ds, \tag{3.189}$$

and then

$$w_1(\xi) = c_1 \int_0^{\xi} e^{-\frac{2}{3} s^{3/2}} ds \tag{3.190}$$
$$+ \int_0^{\xi} e^{-\frac{2}{3} s^{3/2}} \int_0^{\xi} e^{\frac{2}{3} t^{3/2}} \left(-\frac{2}{\sqrt{t}} \frac{dw_0}{d\xi} + \frac{1}{2} \frac{w_0}{(\sqrt{t})^3} \right) dt \, ds$$

for constants c_0 and c_1. When we write

$$w_0(\xi) = c_0 \left(\int_0^{\infty} e^{-\frac{2}{3} s^{3/2}} ds - \int_{\xi}^{\infty} e^{-\frac{2}{3} s^{3/2}} ds \right)$$

and apply the crude matching condition that

$$\lim_{\xi \to \infty} w_0(\xi) = \lim_{x \to 0} Z_0(x),$$

we determine the unusual constant,

$$c_0 = \frac{1}{\int_0^{\infty} e^{-\frac{2}{3} s^{3/2}} ds} \equiv \left(\frac{3}{2} \right)^{1/3} \frac{1}{\Gamma(2/3)}. \tag{3.191}$$

The asymptotic behavior of w_0 as $\xi \to \infty$ follows by using repeated integrations by parts, i.e.

$$
w_0(\xi) = 1 - c_0 \int_\xi^\infty e^{-\frac{2}{3}s^{3/2}} ds
$$

$$
= 1 - c_0 \left[\frac{1}{\sqrt{\xi}} - \frac{1}{2\xi^2} + \frac{1}{(\sqrt{\xi})^7} + O\left(\frac{1}{\xi^5}\right) \right] e^{-\frac{2}{3}\xi^{3/2}}. \qquad (3.192)
$$

Next, since $w_0 \to 1$ and $\frac{dw_0}{d\xi} \to 0$ as $\xi \to \infty$, w_1 must satisfy $\frac{d^2 w_1}{d\xi^2} + \sqrt{\xi}\frac{dw_1}{d\xi} \sim \frac{1}{2(\sqrt{\xi})^3}$ nearby, so upon integrating, we get

$$
w_1(\xi) \sim K_1 + \int_\xi^\infty e^{-\frac{2}{3}s^{3/2}} \int_s^\infty e^{\frac{2}{3}t^{3/2}} \frac{dt}{2(\sqrt{t})^3} ds
$$
$$
\sim K_1 - \frac{1}{2\xi} - \frac{2}{5(\sqrt{\xi})^5} - \frac{7}{8\xi^4} - \frac{35}{11(\sqrt{\xi})^{11}} + \cdots. \qquad (3.193)
$$

(This could be more simply determined by directly introducing a power series for the limiting behavior of w_1 using undetermined coefficients.) Further, integration by parts implies that

$$
\int_\xi^\infty e^{-\frac{2}{3}s^{3/2}} \int_s^\infty \frac{e^{\frac{2}{3}t^{3/2}}}{2t^{3/2}} dt\, ds \sim -\int_\xi^\infty \frac{ds}{2s^2} = -\frac{1}{2\xi}. \qquad (3.194)
$$

To match w_1 at infinity, we need $K_1 = 0$, so $w_1 \to 0$ as $\xi \to \infty$. Thus,

$$
c_1 \int_0^\infty e^{-\frac{2}{3}s^{3/2}} ds =
$$
$$
-\int_0^\infty e^{-\frac{2}{3}s^{3/2}} \int_0^s e^{\frac{2}{3}t^{3/2}} \left(-\frac{2}{\sqrt{t}}\frac{dw_0}{d\xi} + \frac{1}{2}\frac{w_0}{(\sqrt{t})^3} \right) dt\, ds \qquad (3.195)
$$

specifies c_1. Higher-order matching follows analogously. (We will not attempt a uniformly valid composite expansion.) Clearly, matching near turning points is complicated. (It might be an instance calling for the *neutrix calculus* (cf. van der Corput [100]), where infinities are appropriately canceled.) Our procedure for (3.176) might be compared to that of Miller [318] and Johnson [226] who, respectively, consider the linear problem

$$
\epsilon y'' + 12x^{1/3} y' + y = 0, \quad y(0) = 1, \quad y(1) = 1
$$

and the nonlinear problem

$$
\epsilon y'' + \sqrt{x}\, xy' + y^2 = 0, \quad y(0) = 2, \quad y(1) = 1/3.
$$

We now reconsider the nonlinear two-point problem

$$
\epsilon y'' - 2yy' = 0, \quad 0 \le x \le 1 \qquad (3.196)
$$

with prescribed endvalues $y(0)$ and $y(1)$. Wasow [514] cited this as an example of the *capriciousness* of singular perturbations. (In response, Franz and Roos [158] have written about the capriciousness of numerical methods for singular perturbations.) If we integrate once to get $\epsilon y' = y^2 - \alpha$, we can separate variables to provide the general solution

$$y(x, \epsilon) = -\sqrt{\alpha} \tanh\left(\frac{\sqrt{\alpha}}{\epsilon}(x - \beta)\right)$$

$$= -\sqrt{\alpha}\left(\frac{1 - e^{-\frac{2\sqrt{\alpha}}{\epsilon}(x-\beta)}}{1 + e^{-\frac{2\sqrt{\alpha}}{\epsilon}(x-\beta)}}\right) \tag{3.197}$$

for ϵ-dependent constants α and β. (When α is real, we shall take it to be nonnegative.) The boundary conditions require that

$$y(0) = -\sqrt{\alpha}\left(\frac{1 - e^{\frac{2\sqrt{\alpha}\beta}{\epsilon}}}{1 + e^{\frac{2\sqrt{\alpha}\beta}{\epsilon}}}\right) \quad \text{and} \quad y(1) = -\sqrt{\alpha}\left(\frac{1 - e^{-\frac{2\sqrt{\alpha}}{\epsilon}}e^{\frac{2\beta\sqrt{\alpha}}{\epsilon}}}{1 + e^{-\frac{2\sqrt{\alpha}}{\epsilon}}e^{\frac{2\beta\sqrt{\alpha}}{\epsilon}}}\right),$$

so

$$e^{\frac{2\beta\sqrt{\alpha}}{\epsilon}} = \frac{\sqrt{\alpha} + y(0)}{\sqrt{\alpha} - y(0)} = \left(\frac{\sqrt{\alpha} + y(1)}{\sqrt{\alpha} - y(1)}\right)e^{\frac{2\sqrt{\alpha}}{\epsilon}}. \tag{3.198}$$

We will, curiously, find different sorts of limiting behaviors for y in four different portions of the $y(0)$-$y(1)$ plane of boundary values.

(i) On the *half-line* where $y(0) = -y(1) > 0$: Because of the sign of the coefficient of y' in (3.196), we might expect y to be nearly constant near both $x = 0$ and 1. Thus, we can anticipate having the limit

$$y \sim y(0) > 0 \quad \text{near } x = 0$$

and, likewise,

$$y \sim y(1) < 0 \quad \text{near } x = 1.$$

Symmetry even suggests that a narrow *shock* (or *transition*) *layer* between these outer solutions will occur about the midpoint $x = 1/2$ since $y(1) = -y(0)$. Indeed, (3.198) implies that

$$e^{\frac{\sqrt{\alpha}}{\epsilon}} = \frac{\sqrt{\alpha} + y(0)}{\sqrt{\alpha} - y(0)}$$

(corresponding to $\beta = 1/2$) and to the implicit relation

$$\sqrt{\alpha} = y(0) + (\sqrt{\alpha} + y(0))e^{-\sqrt{\alpha}/\epsilon}$$

for α. Iterating, we then find

$$\sqrt{\alpha} \sim y(0) + 2y(0)e^{-y(0)/\epsilon} + \dots \tag{3.199}$$

We recognize this as a result involving exponential asymptotics (i.e., it uses asymptotically negligible correction terms like $e^{-y(0)/\epsilon}$). Thus, the limiting uniform solution

$$y(x, \epsilon) \sim -y(0) \tanh\left(\frac{y(0)}{\epsilon}\left(x - \frac{1}{2}\right)\right) \qquad (3.200)$$

features an $O(\epsilon)$-thick shock layer at the midpoint with the constant limit $y(0)$ for $x < \frac{1}{2}$ and $y(1)$ for $x > 1/2$.

(ii) In that 135° *sector* of the $y(0)$-$y(1)$ plane where $y(0) > 0$ and $y(0) + y(1) > 0$, we might expect the dominant endvalue $y(0)$ to provide the limiting solution, except in a narrow terminal layer near $x = 1$. To see this, rewrite the boundary conditions (3.198) as

$$e^{2\sqrt{\alpha}(\beta-1)/\epsilon} = \frac{\sqrt{\alpha} + y(1)}{\sqrt{\alpha} - y(1)}$$

and

$$\sqrt{\alpha} = y(0) + (\sqrt{\alpha} + y(0))e^{-2\sqrt{\alpha}\beta/\epsilon}.$$

Using the general solution (3.197), $\sqrt{\alpha} \sim y(0)$ implies that

$$y(x, \epsilon) \sim y(0) \left[\frac{y(0) + y(1) - (y(0) - y(1))e^{\frac{2y(0)}{\epsilon}(x-1)}}{y(0) + y(1) + (y(0) - y(1))e^{\frac{2y(0)}{\epsilon}(x-1)}} \right], \qquad (3.201)$$

so y indeed has the constant limit $y(0)$ for $x < 1$ and an ordinary $O(\epsilon)$-thick boundary layer near $x = 1$.

Curiously, when $y(0) + y(1)$ is positive, but only asymptotically exponentially small, the previously found shock wave can be moved all the way from the midpoint $x = 1/2$ to the endpoint $x = 1$. This demonstrates the *supersensitivity* of the shock location β. Imagine the computational consequences!

(iii) One could analogously show (cf. (3.176)) that the limiting solution is $y(1)$, except in an initial layer, when $y(1) < 0$ and $y(0) + y(1) < 0$. Now, for $y(0) + y(1)$ appropriately exponentially negligible, the shock can be moved from $x = \frac{1}{2}$ to $x = 0$. (This also follows by reflection from (ii).)

(iv) In the *quarter-plane* where $y(0) < 0 < y(1)$, we'd expect two endpoint layers. Instead of letting α be imaginary, we take the general solution to have the form

$$y(x, \epsilon) = \epsilon A \tan(A(x - \epsilon B)) \qquad (3.202)$$

with a trivial limit in $0 < x < 1$ and boundary layers at both $x = 0$ and 1.

The limiting possibilities for solutions of (3.196) are illustrated in Fig. 3.1.

The preceding analysis for (3.196) anticipates the corresponding asymptotics for Burgers' partial differential equation

$$u_t = \epsilon u_{xx} + u u_x \tag{3.203}$$

on the planar strip where $-1 \leq x \leq 1$ and $t \geq 0$. For the constant boundary values $u(\pm 1, t) = \pm 1$ and a prescribed smooth initial function $u(0, t)$, we might anticipate the development of a moving shock layer solution

$$\tanh \left(\frac{x - x_\epsilon(t)}{2\epsilon} \right) \tag{3.204}$$

(cf. Reyna and Ward [416] and Laforgue and O'Malley [275]). One can, indeed, use the Cole–Hopf transformation

$$v(\eta, t) = e^{\int_0^\eta u(s,t)\, ds} \quad \text{for } \eta = \frac{x - x_\epsilon}{2\epsilon}$$

to convert Burgers' equation to the linear heat equation and to then solve that using Fourier series. Note the relation to the previously introduced Riccati transformations. One finds that the profile (3.204) moves asymptotically slowly after the shock is formed, according to the equation

$$\frac{dx_\epsilon}{dt} = e^{-1/\epsilon}(e^{-x_\epsilon/\epsilon} - e^{x_\epsilon/\epsilon}). \tag{3.205}$$

The trivial rest point of (3.205) is reached after an asymptotically exponentially long time (i.e., we attain *metastability* due to the asymptotically negligible speed of the shock location x_ϵ). See O'Malley and Ward [379] for study of a variety of related problems. Also note the relationship to *intermediate asymptotics* and self-similar solutions, as presented by Barenblatt [28].

E [130] outlines a two-scale approach to satisfy the *Allen–Cahn equation* (describing phase transitions):

$$u_t = \epsilon \Delta u - \frac{1}{\epsilon} V'(u) \tag{3.206}$$

where V is a double-well potential with local minima u_1 and u_2. He introduces the stretched variable

$$\frac{\varphi(x,t)}{\epsilon}, \tag{3.207}$$

with φ being the distance from x to a boundary curve Γ_t of the domain and makes the multi-scale ansatz

$$u\left(x, t, \frac{\varphi}{\epsilon}, \epsilon\right) = U_0\left(\frac{\varphi(x,t)}{\epsilon}\right) + \epsilon U_1\left(\frac{\varphi(x,t)}{\epsilon}, x, t\right) + \dots \tag{3.208}$$

where $U_0(\pm\infty) = u_{1/2}$. Leading terms in (3.206) then imply that

$$\varphi_t U_0' = U_0'' - V'(U_0)$$

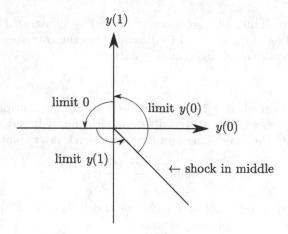

Figure 3.1: The limiting solution to $\epsilon y'' - 2yy' = 0$ differs in four different regions of the $y(0)$-$y(1)$ plane

so

$$\varphi_t = \frac{V(u_1) - V(u_2)}{\int_{-\infty}^{\infty} (U_0'(y))^2 \, dy}. \tag{3.209}$$

When $\dot{V}(u_1) = V(u_2)$, he gets shock layer motion on a longer time scale. Such multi-scale ideas will be developed in Chap. 5.

Cole [92] shows that the asymptotic solution of the boundary value problem

$$\epsilon y'' + yy' - y = 0, \quad 0 \leq x \leq 1, \quad y(0) \text{ and } y(1) \text{ prescribed} \tag{3.210}$$

(less tractable than (3.196)) also varies significantly depending on the end-values $y(0)$ and $y(1)$. This problem was introduced at Caltech in the 1950s as a nonlinear model where solutions feature both endpoint boundary and interior shock layers. It's often called the Cole–Lagerstrom problem. We shall illustrate some possibilities. See Cole [92], Dorr et al. [124], Chang and Howes [76], and Lagerstrom [276] for the complete list of possible asymptotic solutions in nine distinct subsets of the $y(0)$-$y(1)$ plane. Shen and Han [451] provide results for a more general equation.

Note that the reduced equation

$$Y_0(Y_0' - 1) = 0 \tag{3.211}$$

has the trivial solution $Y_0(x) \equiv 0$ and the linear family of solutions $Y_0(x) = x + c$.

(a) Suppose $y(0) = 0$ and $y(1) = 2$. If we take $c = 1$, $Y_0(x) = x + 1$ will satisfy the terminal condition $Y_0(1) = 2$. The positivity of $x + 1$ throughout $0 \leq x \leq 1$ suggests that Y_0 might serve as an outer solution $Y(x, \epsilon)$ for an asymptotic solution with an initial layer, say

$$y(x, \epsilon) = x + 1 + v(\xi, \epsilon) \tag{3.212}$$

where $v(0, \epsilon) = -1$, $v \to 0$ as $\xi = x/\epsilon \to \infty$, and

$$v \sim \sum_{k \geq 0} v_k(\xi) \epsilon^k. \tag{3.213}$$

Then, $y' = 1 + \frac{1}{\epsilon} \frac{dv}{d\xi}$ and $\epsilon y'' = \frac{1}{\epsilon} \frac{d^2 v}{d\xi^2}$ imply that $\frac{d^2 v}{d\xi^2} + (\epsilon\xi + 1 + v)\frac{dv}{d\xi} = 0$, so the leading term v_0 must satisfy the autonomous equation $\frac{d^2 v_0}{d\xi^2} + (1 + v_0)\frac{dv_0}{d\xi} = 0$. Integrating backwards from infinity, we obtain the initial value problem

$$\frac{dv_0}{d\xi} + v_0 + \frac{v_0^2}{2} = 0, \quad v_0(0) = -1. \tag{3.214}$$

Integrating this Riccati equation provides $v_0(\xi) = -\frac{2}{1 + e^\xi}$, i.e., the uniformly valid approximation

$$y(x, \epsilon) \sim x + 1 - \frac{2}{1 + e^{x/\epsilon}}, \tag{3.215}$$

featuring an initial layer of $O(\epsilon)$-thickness in x.

(b) If we instead take $y(0) = -1$ and $y(1) = 1$, we might anticipate having an interior shock layer between the linear left- and right-sided outer solutions

$$Y_L(x) = x - 1 \quad \text{and} \quad Y_R(x) = x. \tag{3.216}$$

(Since $Y_L < 0$ and $Y_R > 0$, we wouldn't expect endpoint layers.) Thus, we will assume an asymptotic solution of the form

$$y(x, \epsilon) = x - 1 + u(\kappa, \epsilon) \tag{3.217}$$

for the stretched variable

$$\kappa \equiv \frac{x - \tilde{x}}{\epsilon}, \tag{3.218}$$

expecting a monotonic unit jump in y about the shock location $\tilde{x}$ (to be determined) such that

$$u \to \begin{cases} 0 & \text{as } \kappa \to -\infty \\ \tilde{x} - (\tilde{x} - 1) = 1 & \text{as } \kappa \to \infty. \end{cases}$$

Since $y' = 1 + \frac{1}{\epsilon}\frac{du}{d\kappa}$ and $\epsilon y'' = \frac{1}{\epsilon}\frac{d^2u}{d\kappa^2}$, u must satisfy $\frac{d^2u}{d\kappa^2} + (\tilde{x} + \epsilon\kappa - 1 + u)\frac{du}{d\kappa} = 0$. Its leading term will satisfy

$$\frac{du_0}{d\kappa} + (\tilde{x} - 1)u_0 + \frac{u_0^2}{2} = 0, \qquad (3.219)$$

upon integrating from $-\infty$. The rest points are 0 and $2(1 - \tilde{x})$. To get a solution joining the rest points $u_0(-\infty) = 0$ and $u_0(\infty) = 1$ requires taking

$$\tilde{x} = 1/2, \qquad (3.220)$$

as we should have anticipated from symmetry. The corresponding limiting shock layer solution is

$$u_0(\kappa) = \frac{1}{1 + e^{-\kappa/\epsilon}} \quad \text{for } \kappa = \frac{x - \frac{1}{2}}{\epsilon}. \qquad (3.221)$$

With analogous nonsymmetric boundary values, the jump would instead be located elsewhere. Higher-order terms follow readily.

Example

The two-point problem

$$\epsilon y'' = 1 - (y')^2, \quad 0 \leq x \leq 1, \quad \text{with} \quad y(0) = 0 = y(1) \qquad (3.222)$$

can be solved by converting it to the slow-fast system

$$\begin{cases} y' = z \\ \epsilon z' = 1 - z^2. \end{cases} \qquad (3.223)$$

If we select the left and right outer solutions

$$Z_L(x) = -1 \quad \text{and} \quad Z_R(x) = 1,$$

corresponding to

$$Y_L(x) = -x \quad \text{and} \quad Y_R(x) = x - 1,$$

the two outer solutions Y_L and Y_R meet at $x = 1/2$, where y' must jump.

We naturally look for a shock layer as a function of the stretched variable

$$\xi = \frac{1}{\epsilon}\left(x - \frac{1}{2}\right).$$

A direct integration (with the "right" endvalues) provides

$$\begin{cases} z(x,\epsilon) = \tanh((x - \frac{1}{2})/\epsilon) \\ \text{and} \\ y(x,\epsilon) = \epsilon \ln\left(\frac{\cosh((x-\frac{1}{2})/\epsilon)}{\cosh(1/2\epsilon)}\right) \end{cases} \qquad (3.224)$$

with the anticipated *angular asymptotics*. Note that y has a minimum at $x = 1/2$. (A maximum principle argument would rule out the selection $Z_L(x) = 1$, $Z_R(x) = -1$.)

Exercises

1. Numerically solve
$$\epsilon y'' = 2yy', \quad 0 \le x \le 1$$
with various boundary values $y(0)$ and $y(1)$ to illustrate the four possible types of limiting solution.

2. For $y(0) = -1/4$ and $y(1) = 1/2$, show that the limiting (piecewise linear) solution to the Cole–Lagerstrom equation (3.210) is given by
$$y(x, \epsilon) \to \begin{cases} x - \frac{1}{4}, & 0 \le x \le \frac{1}{4} \\ 0, & \frac{1}{4} \le x \le \frac{1}{2} \\ x - \frac{1}{2}, & \frac{1}{2} \le x \le 1. \end{cases}$$

3. Show that the nonlinear boundary value problem
$$\epsilon y'' + f(y)y' + g(y) = 0$$
with $y(0)$ and $y(1)$ prescribed can be converted to a slow-fast problem in the y-z (or *Liénard*) *plane* with $z \equiv \epsilon y' + \int^y f(r)\, dr$.

For the more general nonlinear scalar problem
$$\begin{cases} \epsilon y'' + f(x, y)y' + g(x, y) = 0, & 0 \le x \le 1 \\ \text{with } y(0) \text{ and } y(1) \text{ prescribed,} \end{cases} \qquad (3.225)$$
perhaps first studied in Coddington and Levinson [90], we will seek an initial layer solution in the composite form
$$y(x, \epsilon) = Y(x, \epsilon) + v(\xi, \epsilon) \qquad (3.226)$$
where $v \to 0$ as $\xi = x/\epsilon \to \infty$. We will naturally require two stability assumptions:

(i) that the reduced problem
$$f(x, Y_0)Y_0' + g(x, Y_0) = 0, \quad Y_0(1) = y(1) \qquad (3.227)$$
has a solution $Y_0(x)$ on $0 \le x \le 1$ with
$$f(x, Y_0(x)) > 0 \qquad (3.228)$$

(ii) that the separable limiting integrated initial layer problem

$$\begin{cases} \frac{dv_0}{d\xi} + \int_0^{v_0} f(0, Y_0(0) + r)\, dr = 0, \\ v_0(0) = y(0) - Y_0(0) \end{cases} \qquad (3.229)$$

has a solution $v_0(\xi)$ defined throughout $\xi \geq 0$ that decays to zero as $\xi \to \infty$.

Clearly, the outer solution $Y(x, \epsilon) \sim \sum_{j \geq 0} Y_j(x)\epsilon^j$ must satisfy the terminal value problem

$$\epsilon Y'' + f(x, Y)Y' + g(x, Y) = 0, \quad Y(1, \epsilon) = y(1). \qquad (3.230)$$

Since we have assumed existence of the $\epsilon = 0$ solution $Y_0(x)$ throughout $0 \leq x \leq 1$, the next term Y_1 must satisfy the linearized problem

$$f(x, Y_0)Y_1' + \big(f_y(x, Y_0)Y_0' + g_y(x, Y_0)\big)Y_1 + Y_0'' = 0, \quad Y_1(1) = 0. \qquad (3.231)$$

Presuming smoothness of the coefficients, it is no problem to successively define all the Y_ks in terms of the attractive outer limit Y_0.

Knowing $Y(x, \epsilon)$ asymptotically, the supplementary initial layer correction v must satisfy

$$\left(\epsilon Y'' + \frac{1}{\epsilon}\frac{d^2 v}{d\xi^2}\right) + f(x, Y + v)\left(Y' + \frac{1}{\epsilon}\frac{dv}{d\xi}\right) + g(x, Y + v) = 0$$

and $v(0, \epsilon) = y(0) - Y(0, \epsilon)$, i.e.

$$\begin{aligned} \frac{d^2 v}{d\xi^2} &+ f(x, Y + v)\frac{dv}{d\xi} \\ &+ \epsilon\big[(f(x, Y + v) - f(x, Y))Y' + g(x, Y + v) - g(x, Y)\big] = 0 \end{aligned} \qquad (3.232)$$

when we substitute $-f(x, Y)Y' - g(x, Y)$ for $\epsilon Y''$ and expand all functions of $x = \epsilon\xi$ in Taylor series about $x = 0$. Thus, v_0 must be a decaying solution of the nonlinear terminal value problem

$$\frac{d^2 v_0}{d\xi^2} + f(0, Y_0(0) + v_0)\frac{dv_0}{d\xi} = 0, \quad v_0(\infty) = 0. \qquad (3.233)$$

Integrating backwards, we require v_0 to satisfy the initial value problem (3.229). Existence of v_0 is guaranteed by the second stability condition. The next terms in (3.232) require v_1 to satisfy

$$\begin{aligned} \frac{d^2 v_1}{d\xi^2} &+ f(0, Y_0(0) + v_0)\frac{dv_1}{d\xi} + \xi\big[f_x(0, Y_0(0) + v_0) \\ &+ f_y(0, Y_0(0) + v_0)Y_0'(0)\big]\frac{dv_0}{d\xi} \\ &+ \Big[(f(0, Y_0(0) + v_0) - f(0, Y_0(0)))Y_0'(0) \\ &+ (g(0, Y_0(0) + v_0) - g(0, Y_0(0)))\Big] = 0. \end{aligned} \qquad (3.234)$$

Because v_0 and $\frac{dv_0}{d\xi}$ decay exponentially to zero as $\xi \to \infty$, we can integrate backwards from infinity where $v_1(\infty) = 0$. Then, we integrate the resulting linear initial value problem with $v_1(0) = -Y_1(0)$ to get $v_1(\xi)$. Later terms follow analogously, in turn. We note that more direct multi-scale methods for (3.225) (which we will later consider) do not seem to be generally available except when $f(x, y)$ is independent of y.

Example

Consider

$$\epsilon y'' + e^y y' = 1, \quad y(0) = 0, \quad y(1) = 1. \tag{3.235}$$

The limiting outer problem

$$e^{Y_0} Y_0' = 1, \quad Y_0(1) = 1 \tag{3.236}$$

implies that $e^{Y_0} = x + c$ where $e = 1 + c$, so

$$Y_0(x) = \ln(x + e - 1) \tag{3.237}$$

and $e^{Y_0} > 0$. We will seek a uniform limit

$$y(x, \epsilon) \sim Y_0(x) + v_0(\xi) \text{ for } \xi = x/\epsilon. \tag{3.238}$$

Then, v_0 must satisfy $\frac{d^2 v_0}{d\xi^2} + e^{Y_0(0)+v_0(\xi)} \frac{dv_0}{d\xi} = 0$, $Y_0(0) + v_0(0) = 0$ and $v_0 \to 0$ as $\xi \to \infty$. An integration requires v_0 to satisfy the nonlinear initial value problem

$$\frac{dv_0}{d\xi} + (e - 1)(e^{v_0} - 1) = 0, \quad v_0(0) = -\ln(e - 1).$$

Thus

$$v_0(\xi) = -\ln(1 + (e - 2)e^{-(e-1)\xi}) \tag{3.239}$$

and the uniformly valid limiting solution on $0 \le x \le 1$ is

$$y(x, \epsilon) = \ln\left(\frac{x + e - 1}{1 + (e - 2)e^{-(e-1)x/\epsilon}}\right) + O(\epsilon). \tag{3.240}$$

We will next outline the construction of the asymptotic solution of the scalar two-point problem

$$\epsilon y'' + f(x, y)y' + g(x, y) = 0 \text{ with } y(0) \text{ and } y(1) \text{ prescribed} \tag{3.241}$$

featuring a sharp *transition* (i.e., a shock) *layer* at an interior point $\tilde{x}$. (Recall several examples considered previously.)

We will *assume*

(i) that the left limiting problem

$$f(x, Y)Y' + g(x, Y) = 0, \quad Y(0) = y(0) \tag{3.242}$$

has a solution $Y_{L0}(x)$ such that

$$f(x, Y_{L0}(x)) < 0 \quad \text{for } 0 \le x \le \tilde{x}, \tag{3.243}$$

and

(ii) the right limiting problem

$$f(x, Y)Y' + g(x, Y) = 0, \quad Y(1) = y(1) \tag{3.244}$$

has a solution $Y_{R0}(x)$ such that

$$f(x, Y_{R0}(x)) > 0, \quad \text{for } \tilde{x} \le x \le 1. \tag{3.245}$$

Here, the *isolated* jump location $\tilde{x}$ is determined by the classical *Rankine–Hugoniot* jump condition

$$\int_{Y_{L0}(\tilde{x})}^{Y_{R0}(\tilde{x})} f(\tilde{x}, s)\, ds = 0 \tag{3.246}$$

(cf. Whitham [520]).

We will also *assume* that

(iii) the integrated shock layer problem

$$\frac{du_0}{d\kappa} + \int_0^{u_0} f\left(\tilde{x}, Y_{L0}(\tilde{x}) + r\right) dr = 0, \quad -\infty < \kappa < \infty \tag{3.247}$$

has a monotonic solution $u_0(\kappa)$ on $-\infty < \kappa < \infty$ satisfying

$$u_0(-\infty) = 0 \quad \text{and} \quad u_0(\infty) = Y_{R0}(\tilde{x}) - Y_{L0}(\tilde{x}).$$

Then, we can construct an asymptotic solution to (3.241) of the form

$$y(x, \epsilon) = Y_L(x, \epsilon) + u(\kappa, \epsilon) \tag{3.248}$$

where

$$\kappa = \frac{1}{\epsilon}(x - \tilde{x}), \tag{3.249}$$

$$u(-\infty, \epsilon) = 0, \quad \text{and} \quad u(\infty, \epsilon) \to Y_R(\tilde{x} + \epsilon\kappa, \epsilon) - Y_L(\tilde{x} + \epsilon\kappa, \epsilon)$$

for left- and right-outer expansions $Y_L(x, \epsilon)$ and $Y_R(x, \epsilon)$ with limits Y_{L0} and Y_{R0} such that

$$Y_L(x, \epsilon) \sim \sum_{j \ge 0} Y_{Lj}(x)\epsilon^j, \quad Y_R(x, \epsilon) \sim \sum_{j \ge 0} Y_{Rj}(x)\epsilon^j,$$

$$\text{and} \quad u(\kappa, \epsilon) \sim \sum_{j \ge 0} u_j(\kappa)\epsilon^j. \tag{3.250}$$

As an alternative, we could center the shock layer at the average value

$$\frac{1}{2}\left(Y_L(\tilde{x}, \epsilon) + Y_R(\tilde{x}, \epsilon)\right)$$

and seek a shock layer solution $v(\kappa, \epsilon)$ tending to $Y_L(\tilde{x}, \epsilon)$ as $\kappa \to -\infty$ and to $Y_R(x, \epsilon)$ as $\kappa \to +\infty$.

Lorenz [299] considered the example

$$\begin{cases} \epsilon y'' + y(1 - y^2)y' - y = 0 \\ y(0) = 1.6, \quad y(1) = -1.7. \end{cases}$$

He showed that the limiting solution satisfies

$$y(x, \epsilon) \to \begin{cases} Y_{L0}(x), & 0 < x < x_1 = \frac{\sqrt{2}}{3} - 1.6 - \frac{(1.6)^3}{3} \\ 0, & x_1 < x < x_2 \\ Y_{R0}(x), & -\frac{\sqrt{2}}{3} + 1.7 - \frac{(1.7)^3}{3} = x_2 < x \le 1 \end{cases}$$

for left and right limiting solutions $Y_{L0}(x)$ and $Y_{R0}(x)$ and with $x_1 \approx 0.24$ and $x_2 \approx 0.59$, found computationally. Readers should verify this conclusion analytically or numerically.

We might expect solutions of two-point problems for the semilinear vector equation

$$\epsilon^2 y'' + f(x, y) = 0, \quad 0 \le x \le 1 \tag{3.251}$$

to instead converge away from boundary layers at both endpoints to a solution Y_0 of the limiting algebraic equation

$$f(x, Y_0(x)) = 0 \tag{3.252}$$

when (i) the Jacobian

$$f_y(x, Y_0(x))$$

is a *stable* matrix throughout the interval and (ii) the corresponding boundary layer jumps $y(0) - Y_0(0)$ and $y(1) - Y_0(1)$ are appropriately restricted to achieve stability in the boundary layers.

Then, the asymptotic solution to (3.251) will take the form

$$y(x, \epsilon) = Y(x, \epsilon) + r(\kappa, \epsilon) + s(\lambda, \epsilon) \tag{3.253}$$

where the terms of $r \to 0$ as $\kappa = \frac{x}{\epsilon} \to \infty$ while $s \to 0$ as $\lambda = \frac{1-x}{\epsilon} \to \infty$. Franz and Roos [158], likewise, show that the limiting solution to

$$\epsilon^2 y'' - y(y-1)\left(y - x - \frac{3}{2}\right) = 0, \quad y(0) = 0, \quad y(1) = 5/2$$

satisfies

$$y(x, \epsilon) \to \begin{cases} 0, & 0 \le x < \frac{1}{2} \\ x + \frac{3}{2}, & \frac{1}{2} < x \le 1. \end{cases}$$

These boundary values don't require endpoint layers. However, a shock layer between the two outer limits is needed at $x = 1/2$.

A semilinear example with (at least) two solutions having both boundary and corner layers is contained in O'Donnell [356]. It is the system

$$\begin{cases} \epsilon y_1'' = \left(y_1 - \left|x - \tfrac{1}{2}\right|\right)\left(1 + y_2^2\right) \\ \epsilon y_2'' = \left(y_2 - 1 + \left|\tfrac{1}{3} - x\right|\right)\left(1 + y_1^2\right) \end{cases}$$

for appropriate values $y_1(0)$, $y_1(1)$, $y_2(0)$, and $y_2(1)$.

Other two-point problems are considered by Butuzov et al. [64]. Indeed, Butuzov et al. [63] and Vasil'eva et al. [497] develop a theory for the so-called *contrast structures*. See Schmeiser [439] for an application with ϵ-dependent boundary conditions. Somewhat comparable asymptotic results to those we've obtained are also available for certain integral equations (cf. Shubin [454] and its references) and other functional equations.

Overview (or Proof of Asymptotic Validity)

When one solves boundary value problems asymptotically, one typically uses only a few terms in the formal series generated, because of the effort involved and because the series generally diverge. Thus, if one knows the first two terms $y_0(x, \eta)$ and $y_1(x, \eta)$ in a formal approximation, with the given independent variable x and, say, a stretched variable $\eta(x, \epsilon)$ describing a boundary or interior layer of nonuniform convergence, we might write the actual solution $y(x, \epsilon)$ as

$$y(x, \epsilon) = y_0(x, \eta) + \epsilon y_1(x, \eta) + \epsilon^2 R(x, \epsilon) \qquad (3.254)$$

and convert the given boundary value problem for y into a new problem for the (scaled) remainder R. Often, we further convert the latter problem into an integral equation for R. If we can estimate its solution R using, for example, differential inequalities, the boundedness of R throughout the x interval will imply that our formal result

$$y \sim y_0 + \epsilon y_1$$

is asymptotically correct to $O(\epsilon^2)$ in that interval. Such proofs are given in Smith [466], Murdock [335], and de Jager and Jiang [224].

(d)   Linear Boundary Value Problems

So far, we may have casually given the incorrect impression that singularly perturbed boundary value problems have unique asymptotic solutions, typically consisting of an outer solution and endpoint boundary layer corrections.

The actual situation is illustrated quite clearly by the constant linear n-dimensional vector system

$$\epsilon y' = Ay, \quad 0 \le x \le 1 \tag{3.255}$$

subject to n coupled linear boundary conditions

$$\alpha y(0) + \beta y(1) = \gamma \tag{3.256}$$

for scalar constants α and β and an n-vector γ. More complications naturally occur when the entries of the matrix A vary with x.

We will assume here that the matrix A has the spectral decomposition

$$A = PDP^{-1} \tag{3.257}$$

for a nonsingular constant matrix P and a block-diagonal matrix

$$D \equiv \text{diag } (S, 0, U) \tag{3.258}$$

where S is a stable $r \times r$ matrix, 0 is the trivial $s \times s$ matrix, and U is a $t \times t$ unstable matrix with $r + s + t = n$. Then the so-called *shearing* transformation

$$y = Pz \tag{3.259}$$

implies the two-point problem

$$\epsilon z' = Dz, \quad \alpha Pz(0) + \beta Pz(1) = \gamma \tag{3.260}$$

for z. If we split

$$z \equiv \begin{pmatrix} z_1 \\ z_2 \\ z_3 \end{pmatrix} \tag{3.261}$$

for an r-dimensional vector z_1, an s-dimensional z_2, and a t-dimensional z_3, bounded solutions z must be of the form

$$\begin{cases} z_1(x) = e^{Sx/\epsilon} z_1(0), \\ z_2(x) = z_2(0), \\ \text{and} \\ z_3(x) = e^{-U(1-x)/\epsilon} z_3(1) \end{cases} \tag{3.262}$$

for bounded endvalues $z_1(0)$, $z_2(0)$, and $z_3(1)$ that act like *shooting* parameters in the transformed linear boundary condition

$$\alpha P \begin{pmatrix} z_1(0) \\ z_2(0) \\ e^{-U/\epsilon} z_3(1) \end{pmatrix} + \beta P \begin{pmatrix} e^{S/\epsilon} z_1(0) \\ z_2(0) \\ z_3(1) \end{pmatrix} = \gamma.$$

Note that the matrix entries $e^{-U/\epsilon}$ and $e^{S/\epsilon}$ are asymptotically negligible. Asymptotically, then, we obtain a unique solution

$$y = P \begin{pmatrix} e^{Sx/\epsilon} z_1(0) \\ z_2(0) \\ e^{-\frac{U}{\epsilon}(1-x)} z_3(1) \end{pmatrix} \tag{3.263}$$

of the given boundary value problem when we can uniquely solve the limiting n-dimensional linear system

$$\alpha P \begin{pmatrix} z_1(0) \\ z_2(0) \\ 0 \end{pmatrix} + \beta P \begin{pmatrix} 0 \\ z_2(0) \\ z_3(1) \end{pmatrix} \sim \gamma \tag{3.264}$$

for the shooting vectors $z_1(0)$, $z_2(0)$, and $z_3(1)$. Note that the resulting solution (3.263) features an r-dimensional initial layer determined by $z_1(0)$, an s-dimensional constant outer solution determined by $z_2(0)$, and a t-dimensional terminal layer determined by $z_3(1)$. If the Jacobian of (3.264) (with respect to these n unknown endvalues) is singular, however, there may be multiple solutions of the problem (3.255)–(3.256) or none at all. Generalizations to block-diagonalizable slow-fast linear systems without turning points are straightforward (cf. Harris [197], Flaherty and O'Malley [148], and O'Malley [368]). Surveys of classical results are contained in Wasow [512, 513], Hsieh and Sibuya [220], and Balser [25].

Wasow [511, 517] considered the higher-order variable coefficient linear scalar equation

$$\epsilon^{\ell-k} L(y) + K(y) = 0, \quad 0 \le x \le 1 \tag{3.265}$$

where $L(y)$ is an ℓth-order linear differential operator with leading term

$$y^{(\ell)} \tag{3.266}$$

and where $K(y)$ is a k-th order linear differential operator with leading term

$$\beta_0(x) y^{(k)} \tag{3.267}$$

for $\ell > k \ge 0$ and with $\beta_0(x) \ne 0$ throughout the interval, thereby avoiding turning points. The prototype differential equation is

$$\epsilon^{\ell-k} y^{(\ell)} + \beta_0(x) y^{(k)} = 0.$$

He also prescribed r linear scalar initial conditions

$$A_i y(0) = \gamma_i \tag{3.268}$$

for (3.265) with

$$A_i y = y^{(\lambda_i)} + \text{lower-order terms}, \quad i = 1, \dots, r \tag{3.269}$$

for decreasing orders λ_i as well as s linear terminal conditions

$$A_j y(1) = \gamma_j \tag{3.270}$$

with

$$A_j y = y^{(\lambda_j)} + \text{lower-order terms}, \quad j = r+1, \ldots, r+s = \ell \tag{3.271}$$

for decreasing λ_js. (Treating boundary conditions coupling derivatives at the two endpoints would again be more complicated.)

Assuming appropriate smoothness of the coefficients, one can construct a complete set of ℓ smooth linearly independent asymptotic solutions of (3.265) of the form

$$\begin{cases} G_j(x, \epsilon) e^{\frac{1}{\epsilon} \int^x (-\beta_0(s))^{1/(\ell-k)} ds}, & j = 1, 2, \ldots, \ell - k \\ \text{and} & \\ G_j(x, \epsilon), & j = \ell - k + 1, \ldots, \ell \end{cases} \tag{3.272}$$

for power series G_j to be determined and distinct roots $(-\beta_0(x))^{\frac{1}{\ell-k}}$, generalizing our WKB results for $\ell = 2$ and $k = 1$. (Note the fixed regular spacing of these roots in the complex plane.) The first $\ell - k$ of these solutions display boundary layer behavior near $x = 0$ whenever

$$\text{Re } (-\beta_0(x))^{\frac{1}{(\ell-k)}} < 0.$$

We will, more explicitly, use the representation

$$G_j(x, \epsilon) e^{\frac{1}{\epsilon} \int_0^x (-\beta_0(s))^{\frac{1}{\ell-k}} ds} \tag{3.273}$$

to specify initial layer behavior. Likewise, we will use solutions

$$G_j(x, \epsilon) e^{\frac{1}{\epsilon} \int_x^1 (-\beta_0(s))^{\frac{1}{\ell-k}} ds} \tag{3.274}$$

with terminal layers near $x = 1$ when

$$\text{Re } (-\beta_0(x))^{\frac{1}{\ell-k}} > 0$$

holds. We suppose that

$$\sigma \tag{3.275}$$

of the ℓ-k values $\text{Re } (-\beta_0(x))^{\frac{1}{\ell-k}}$ are negative, that

$$\tau \tag{3.276}$$

are positive, and that we are in the nonexceptional case when

$$\sigma + \tau = \ell - k. \tag{3.277}$$

(Then, none of the roots are purely imaginary.) The last k asymptotic solutions (3.272) don't feature boundary layer behavior and they can, indeed, be found as regular perturbations of any set of linearly independent solutions of the reduced equation $K(y) = 0$. (Solutions of the corresponding nonhomogeneous equation

$$\epsilon^{\ell-k} L(y) + K(y) = f(x)$$

could be obtained from the ℓ asymptotic solutions (3.272) of the homogeneous equation by using variation of parameters.)

Note that even the harmless-looking two-point problem

$$\epsilon^2 y'' + y = 0, \quad 0 \le x \le 1, \qquad y(0) = 0, \quad y(1) = 1$$

hasn't a limiting solution as $\epsilon \to 0$. The solution

$$y(x, \epsilon) = \frac{\sin \frac{x}{\epsilon}}{\sin \frac{1}{\epsilon}}$$

isn't even defined for $\epsilon = \frac{1}{n\pi}$, $n = 1, 2, \ldots$, and it is rapidly oscillating otherwise. Thus, Wasow's quest wasn't trivial.

Writing the solution of the boundary value problem (3.265–3.271) as a linear combination of the ℓ asymptotic solutions, we get a unique solution only if the appropriate $\ell \times \ell$ determinant obtained from applying the prescribed boundary conditions (3.268) and (3.270) is nonsingular for small values of ϵ. Because of the special form of the differential equation and of the boundary conditions, many entries of the determinant involve a limiting *Vandermonde* form (cf. Horn and Johnson [215]), which allows us to asymptotically factor the determinant conveniently. Indeed, it will often allow us to define a *cancellation law*, implying which k limiting boundary conditions, together with the limiting equation $K(y) = 0$, will uniquely specify the limiting solution $Y(x, 0)$ within $0 < x < 1$. With no purely imaginary roots

$$(-\beta_0(x))^{\frac{1}{\ell-k}},$$

such arguments show that the reduced problem will consist of the reduced equation $K(y) = 0$, the last $r-\sigma$ limiting initial conditions (3.268) (presuming $r \ge \sigma$) and of the last $s - \tau$ limiting terminal conditions (3.270) (presuming $s \ge \tau$). (Recall that

$$(r - \sigma) + (s - \tau) = \ell - (\ell - k) = k,$$

the order of reduced operator K.) The general result involves more complicated algebra, but the approach to take is clear in principle. See Wasow [513] for more details.

Wasow's study can be motivated by the simpler question of finding the asymptotic behavior of the ℓ roots $m(\epsilon)$ to the polynomial equation

$$\epsilon^{\ell-k} L(m) + K(m) = 0 \tag{3.278}$$

where K is a polynomial of degree k and L, a polynomial of degree $\ell > k$ (cf. Lin and Segel [291] and Murdock [335]). Indeed, one can also consider the polynomial

$$f(y, \epsilon) = f_\ell(\epsilon)y^\ell + f_{\ell-1}(\epsilon)y^{\ell-1} + \ldots + f_1(\epsilon)y + f_0(\epsilon) = 0. \qquad (3.279)$$

If asymptotic expansions

$$f_s(\epsilon) \sim f_{s0}\epsilon^{\rho_s} + \ldots, \qquad f_{s0} \neq 0 \text{ and } \rho_s \geq 0 \qquad (3.280)$$

are given for the coefficients, it would be reasonable to use a *dominant balance* argument to seek asymptotic solutions satisfying

$$y \sim \epsilon^{\nu_p} y_p, \qquad y_p \neq 0 \qquad (3.281)$$

that provide a limiting balance in (3.279). This is the basis of the *Newton polygon* method. Knowing the limiting approximations (3.281) for the roots, we can readily improve upon them.

To solve problems (3.265–3.271) asymptotically, we don't need the cumbersome machinery of matched asymptotic expansions. Indeed, the advances made by the distinguished American lineage, G. D. Birkhoff, R. E. Langer, H. L. Turrittin, and W. A. Harris, among others, between 1908 and 1960 suffice for such problems, until we encounter turning points or nonlinearities (cf. Turrittin [488] and Schissel [435]). They simply construct, algorithmically, a full linearly independent set of asymptotic solutions. We observe that they were most likely completely unaware of Prandtl's boundary layer theory, though they knew the work of the Germans Fuchs and Frobenius in the nineteenth century and more recent developments by pure mathematicians worldwide. In quite a different direction, Devaney [117] considers complex maps of the form $P(z) + \frac{\lambda}{(z-a)^d}$ for polynomials P, $d > 0$, and λ small.

We note that one way to obtain high-order singularly perturbed differential equations like (3.265) is to consider initial function problems for *delay equations*

$$\dot{x}(t) = f(x(t), \, x(t - \tau))$$

for small values of the delay $\tau > 0$. In particular, if one expands f to some finite order in powers of τ, the highest time derivative occurring will be multiplied by the corresponding power of τ. The resulting long-term solution behavior of the original delay equation and of the approximating differential equation (under suitable hypotheses) can be expected to agree (cf. Chicone [88] and Erneux [143]).

Examples

1. Consider the singularly perturbed problem

$$\epsilon^2 y'''' - y'' = 0, \qquad 0 \leq x \leq 1 \qquad (3.282)$$

with prescribed boundary values

$$y'''(0), \quad y(0), \quad y'(1), \quad \text{and } y(1). \tag{3.283}$$

Linearly independent solutions of the differential equation are given by

$$e^{-x/\epsilon}, e^{x/\epsilon}, \quad 1, \text{ and } x$$

(as can be immediately verified), so we naturally seek a solution of the boundary value problem in the form

$$y(x, \epsilon) = a(\epsilon) + b(\epsilon)x + \epsilon^3 c(\epsilon)e^{-x/\epsilon} + \epsilon d(\epsilon)e^{-(1-x)/\epsilon} \tag{3.284}$$

for constants a, b, c, and d to be asymptotically determined as power series in ϵ (for scale factors ϵ^3 and $\epsilon e^{-1/\epsilon}$ introduced to simplify later algebra). Formulas for the derivatives of y follow directly. The boundary conditions (omitting only asymptotically negligible coefficients like $\frac{e^{-1/\epsilon}}{\epsilon^2}$) imply that

$$\begin{cases} y'''(0) \sim -c(\epsilon) \\ y(0) \sim a(\epsilon) + \epsilon^3 c(\epsilon) \\ y'(1) \sim b(\epsilon) + d(\epsilon) \\ \text{and} \\ y(1) \sim a(\epsilon) + b(\epsilon) + \epsilon d(\epsilon). \end{cases} \tag{3.285}$$

Solving these linear equations implies that the unique asymptotic solution of our two-point problem (3.282–3.283) has the form

$$\begin{aligned} y(x, \epsilon) \sim &\, [y(0) + \epsilon^2 y'''(0)] \\ &+ \frac{x}{1-\epsilon}[y(1) - \epsilon y'(1) - y(0) - \epsilon^3 y'''(0)] - \epsilon^3 y'''(0)e^{-x/\epsilon} \\ &+ \frac{\epsilon e^{-(1-x)/\epsilon}}{1-\epsilon}[y(0) + \epsilon^3 y'''(0) + y'(1) - y(1)]. \end{aligned} \tag{3.286}$$

The limiting solution

$$Y_0(x) = y(0) + (y(1) - y(0))x \tag{3.287}$$

exactly satisfies the reduced problem

$$Y_0'' = 0, \quad Y_0(0) = y(0), \quad Y_0(1) = y(1), \tag{3.288}$$

so a *cancellation law* applies. y' will converge to Y_0', but nonuniformly at $x = 0$ though not at $x = 1$, while higher derivatives of y are generally algebraically unbounded at both endpoints when $\epsilon \to 0$.

2. Boundary value problems need not have unique solutions. Consider, as an example, the planar slow-fast nonlinear system

$$\begin{cases} \dot{x} & = y \\ \epsilon\dot{y} & = -\tfrac{1}{2}(1+3x^2)y \end{cases} \tag{3.289}$$

on $0 \leq t \leq 1$ with the homogeneous separated boundary conditions

$$x(0,\epsilon) + \epsilon y(0,\epsilon) = 0 \quad \text{and} \quad x(1,\epsilon) = 0. \tag{3.290}$$

The two-point problem certainly has the trivial solution. Because $-\tfrac{1}{2}(1+3x^2) < 0$, we might anticipate having an initial layer. Thus, we naturally seek a nontrivial asymptotic solution of the form

$$\begin{cases} x(t,\epsilon) = X(t,\epsilon) + u(\tau,\epsilon) \\ y(t,\epsilon) = Y(t,\epsilon) + \tfrac{1}{\epsilon}v(\tau,\epsilon) \end{cases} \tag{3.291}$$

with an initial layer correction $\begin{pmatrix} u \\ \tfrac{v}{\epsilon} \end{pmatrix}$ tending to zero as the stretched variable

$$\tau = \frac{t}{\epsilon} \tag{3.292}$$

tends to infinity, thereby anticipating an initial impulse in the fast variable y and nonuniform convergence in the slow. Then, the outer solution $\begin{pmatrix} X \\ Y \end{pmatrix}$ must satisfy the given system

$$\begin{cases} \dot{X} = Y \\ \epsilon\dot{Y} = -\tfrac{1}{2}(1+3X^2)Y \end{cases} \tag{3.293}$$

as a power series in ϵ, together with the terminal condition

$$X(1,\epsilon) = 0. \tag{3.294}$$

A regular perturbation procedure readily implies that this outer expansion is trivial to all orders ϵ^k. Thus, the initial layer correction must satisfy the initial value problem

$$\begin{cases} \frac{du}{d\tau} = v \\ \frac{dv}{d\tau} = -\tfrac{1}{2}(1+3u^2)v \end{cases} \tag{3.295}$$

with

$$u(0,\epsilon) + v(0,\epsilon) = 0 \tag{3.296}$$

and the asymptotic stability condition as $\tau \to \infty$. Since $\frac{dv}{d\tau} = -\tfrac{1}{2}(1 + 3u^2)\frac{du}{d\tau}$, integrating from infinity implies that $-2v = u + u^3$, leaving us the initial value problem

$$\frac{du}{d\tau} = -(1+u^2)\frac{u}{2}, \quad u(0,\epsilon) = u^3(0,\epsilon) \tag{3.297}$$

The three possible initial values are

$$u(0, \epsilon) = 0, \quad 1, \quad \text{and} \quad -1. \tag{3.298}$$

The two resulting *nontrivial* solutions are readily found as solutions of the Bernoulli equation for u to be

$$x(t, \epsilon) = \frac{\pm 1}{\sqrt{2e^{t/\epsilon} - 1}} \tag{3.299}$$

and

$$y(t, \epsilon) = \mp \frac{1}{\epsilon} \frac{e^{t/\epsilon}}{(\sqrt{2e^{t/\epsilon} - 1})^3}. \tag{3.300}$$

3. Smith [466] considers the nonlinear two-point problem for

$$\epsilon^2 \ddot{x} = (x^2 - 1)(x^2 - 4). \tag{3.301}$$

With boundary values $x(0) = x(1) = \frac{1}{2}$, he obtains a solution with the outer limit 2 and another with the outer limit -1 (with endpoint layers). However, with the boundary values $\dot{x}(0) = 0$ and $x(1) = \frac{1}{2}$, he obtains a solution with outer limit 2 and another with outer limit -1 (and terminal layers).

To determine the limiting behavior of solutions to singularly perturbed boundary value problems, it is helpful to know about the possibilities exhibited by a variety of solved examples. In this regard, readers are especially referred to the work of the late Fred Howes (cf., e.g., Howes [217], Chang and Howes [76]) who used many explicit examples to motivate more general results. The subtleties arising suggest that blind computation may often be useless. Other challenging problems can, for example, be found in Smith [466], Carrier [70], Bogaevski and Povzner [50], Hinch [206], Johnson [226], Verhulst [500], Cousteix and Mauss [104], Ablowitz [1], Holmes [209], and Paulsen [387], and in the following.

Exercises

1. Show that the asymptotic solution of

$$\begin{cases} \epsilon^2 y'''' - y'' = x^2, \\ y(0) = 0, \quad y'(0) = 1, \quad y'(1) = 2, \quad \text{and} \quad y''(1) = 3 \end{cases}$$

   is given by

$$\begin{aligned} y(x, \epsilon) = &-\frac{1}{12}(x^4 - 28x) + \frac{4\epsilon}{3}\left(-3x - 1 + e^{-x/\epsilon}\right) \\ &+ \epsilon^2\left(-x^2 + 2x + 4 - 4e^{-x/\epsilon} + 4e^{-(1-x)/\epsilon}\right) \\ &+ O(\epsilon^3). \end{aligned}$$

2. Approximate the eigenvalues $\lambda(\epsilon)$ and the corresponding eigenfunctions $y(x, \epsilon)$ for

$$\begin{cases} \epsilon^2 y'''' - y'' = \lambda y, & 0 \leq x \leq 1 \\ y(0) = y'(0) = y(1) = y'(1) = 0 \end{cases}$$

(cf. Moser [329], Handelman et al [193], and Frank [156]).

3. Show that the following boundary value problems have no limiting solution as $\epsilon \to 0^+$:

   (a) $\epsilon y'' - y' = 0$, $\quad y'(0) = 1$, $\quad y(1) = 0$
   (b) $\epsilon^2 y''' + y' = 0$, $\quad y'(0) = 0$, $\quad y(0) = y'(1) = 1$.

4. Consider the initial value problem

$$\epsilon \dot{y} = A(t)y, t \geq 0, \text{ with } y(0) \text{ given}$$

   when

$$A(t) = U^{-1}(t) \begin{pmatrix} -1 & \eta/\epsilon \\ 0 & -1 \end{pmatrix} U(t)$$

   for

$$U(t) = \begin{pmatrix} \cos t & \sin t \\ -\sin t & \cos t \end{pmatrix}.$$

   Show that solutions for $\eta > 2$ can be unbounded, even through $A(t)$ remains stable. Hint: Solve for $v = Uy$ (cf. Kreiss [263]).

5. Solve

$$\epsilon y'' + xy' = x$$

   on $0 \leq x \leq 1$ with $y(0) = y(1) = 1$ and describe the limiting behavior as $\epsilon \to 0$.

6. Show how one can find an asymptotic solution of the two-point problem

$$\epsilon y'' + (1 + x^2)y' + 2xy = x, \qquad 0 \leq x \leq 1$$

   in the form

$$y(x, \epsilon) = A(x, \epsilon) + e^{-\frac{1}{\epsilon} \int_0^x (1+s^2) \, ds} \left(y(1) - A(1, \epsilon) \right).$$

7. Consider the nonlinear two-point problem

$$\epsilon y'' = y' - (y')^3, \quad y(0) = 0, \quad y(1) = \frac{1}{2}.$$

   (Hint: The equation for $z = y'$ can be integrated explicitly.)

Show that the (angular) limiting solution satisfies

$$y(x, \epsilon) \to \begin{cases} 0, & 0 \le x \le \frac{1}{2} \\ x - \frac{1}{2}, & \frac{1}{2} \le x \le 1. \end{cases}$$

Why couldn't you obtain

$$y(x, \epsilon) \to \begin{cases} -x, & 0 \le x \le \frac{1}{4} \\ x - \frac{1}{2}, & \frac{1}{4} \le x \le 1? \end{cases}$$

Such problems are discussed in Chang and Howes [76] and elsewhere. The original reference is Haber and Levinson [189]. Also, see Vishik and Lyusternik [506].

8. Müller et al. [333] considered the two-point problem

$$\begin{cases} \epsilon y'' = y^3, & 0 < x < 1 \\ y(0) = 1, & y(1) = 2. \end{cases}$$

(a) Obtain the exact solution in terms of *elliptic functions*.

(b) Show that the limiting solution within $0 < x < 1$ is trivial and that $\sqrt{\epsilon}$-thick endpoint layers occur.

9. Consider

$$\epsilon y'' = y(y - x), \quad y(-1) = 0, \quad y(1) = 1.$$

Show that the inverse function $x(y)$ satisfies

$$\epsilon \frac{d^2 x}{dy^2} = y(x - y) \left(\frac{dx}{dy} \right)^3, \quad x(0) = -1, \quad x(1) = 1$$

(cf. Howes [218]).

10. Pokrovskii and Sobolev [395] consider the piecewise linear system

$$\begin{cases} \dot{x} = 1, \\ \epsilon \dot{y} = x + |y|. \end{cases}$$

(a) Determine typical trajectories numerically.

(b) Show that

$$y = \begin{cases} x - \epsilon, & x < \epsilon \\ 2\epsilon e^{(x-\epsilon)/\epsilon} - x - \epsilon, & x \ge \epsilon \end{cases}$$

and

$$y = \begin{cases} -x - \epsilon, & x < -\epsilon \\ 2\epsilon e^{-(x+\epsilon)/\epsilon} + x - \epsilon, & -\epsilon < x < \epsilon\nu \\ \epsilon(1 + \nu)e^{(x-\nu\epsilon)/\epsilon} - x - \epsilon, & \epsilon\nu < \epsilon \end{cases}$$

are *invariant manifolds* where ν is a root of $2e^{-1-\nu} + \nu - 1 = 0$.

Lomov [298] used the scalar initial value problem

$$\epsilon u' + \frac{2u}{1+x^2} = 2\left(\frac{3 + (\tan^{-1} x)^2}{1+x^2}\right), \quad u(0) = 1, \quad x \geq 0 \qquad (3.302)$$

to introduce singular perturbations. The reduced problem has the solution

$$U_0(x) = 3 + (\tan^{-1} x)^2.$$

Let us seek an *outer* expansion

$$U(x, \epsilon) = U_0(x) + \epsilon U_1(x) + \epsilon^2 U_2(x) + \dots \qquad (3.303)$$

Substituting this series into (3.302) requires

$$\epsilon U_0' + \epsilon^2 U_1' + \dots + \frac{2}{1+x^2}(U_0 + \epsilon U_1 + \epsilon^2 U_2 + \dots) = \frac{2}{1+x^2}\left(3 + (\tan^{-1} x)^2\right).$$

The ϵ coefficient implies that $U_0' = -\frac{2U_1}{1+x^2}$, so

$$U_1(x) = \tan^{-1} x.$$

Next the ϵ^2 coefficient implies that $U_1' = -\frac{2U_2}{1+x^2}$, or

$$U_2(x) = \frac{1}{2}.$$

Higher coefficients imply that $U_k = 0$ for $k \geq 3$, so we have found an *exact* outer solution

$$U(x, \epsilon) = 3 + (\tan^{-1} x)^2 + \epsilon \tan^{-1} x + \frac{\epsilon^2}{2} \qquad (3.304)$$

of the differential equation. The homogeneous differential equation has the complementary solution

$$e^{-\frac{2}{\epsilon}\int^x \frac{ds}{1+s^2}} = e^{-\frac{2}{\epsilon}\tan^{-1} x}k,$$

so the *exact* solution of our initial value problem (3.302) is

$$u(x, \epsilon) = U(x, \epsilon) + e^{-\frac{2}{\epsilon}\tan^{-1} x}(1 - U(0, \epsilon)). \qquad (3.305)$$

Note that the second term is an initial layer correction increasing from $-2 - \frac{\epsilon^2}{2}$ when $x = 0$ to 0 as $\frac{\tan^{-1} x}{\epsilon} \to \infty$. It is essential since the outer solution doesn't satisfy the prescribed initial condition. By contrast, the matched expansion solution would have the additive form

$$u(x, \epsilon) = U(x, \epsilon) + v(\xi, \epsilon) \qquad (3.306)$$

where $v \to 0$ as $\xi = x/\epsilon \to \infty$. Since the initial layer correction v must satisfy

$$\frac{dv}{d\xi} + \frac{2}{1 + \epsilon^2 \xi^2} v = 0, \quad v(0, \epsilon) = u(0) - U(0, \epsilon), \qquad (3.307)$$

its leading term v_0 must satisfy $\frac{dv_0}{d\xi} + 2v_0 = 0$ and $v_0(0) = -2$. Thus,

$$v_0(\xi) = -2e^{-2\xi} \qquad (3.308)$$

approximates the exact initial layer correction $-e^{-\frac{2}{\epsilon} \tan^{-1} x} \left(2 + \frac{\epsilon^2}{2}\right)$.

The more general scalar problem

$$\epsilon u' = a(x)u + b(x), \quad x \geq 0 \qquad (3.309)$$

has the exact solution

$$u(x, \epsilon) = e^{\frac{1}{\epsilon} \int_0^x a(s)\, ds} u(0) + \frac{1}{\epsilon} \int_0^x e^{\frac{1}{\epsilon} \int_t^x a(s)\, ds} b(t)\, dt. \qquad (3.310)$$

(as noted earlier). Assuming that

$$a(x) < 0, \qquad (3.311)$$

the homogeneous solution

$$e^{\frac{1}{\epsilon} \int_0^x a(s)\, ds}$$

features nonuniform convergence from 1 to 0 in an $O(\epsilon)$-thick initial layer. Again, it is natural to seek an asymptotic solution of (3.309) in the form

$$u(x, \epsilon) = U(x, \epsilon) + e^{\frac{1}{\epsilon} \int_0^x a(s)\, ds}(1 - U(0, \epsilon)) \qquad (3.312)$$

for an outer expansion

$$U(x, \epsilon) = U_0(x) + \epsilon U_1(x) + \epsilon^2 U_2(x) + \dots. \qquad (3.313)$$

Then U must satisfy (3.309) as a power series in ϵ. Equating coefficients successively, we will need

$$a(x)U_0 + b(x) = 0, \quad a(x)U_1 = U_0', \quad a(x)U_2 = U_1',$$

etc., so we uniquely obtain the expansion

$$U(x, \epsilon) = -\frac{b(x)}{a(x)} - \frac{\epsilon}{a(x)}\left(\frac{b(x)}{a(x)}\right)' - \frac{\epsilon^2}{a(x)}\left(\frac{1}{a(x)}\left(\frac{b(x)}{a(x)}\right)'\right)' + \dots \qquad (3.314)$$

presuming sufficient smoothness of the coefficients a and b. As we'd expect, this also follows from (3.310) using repeated integration by parts. Rewriting

$$
\begin{aligned}
u(x, \epsilon) &= e^{\frac{1}{\epsilon} \int_0^x a(s)\, ds} \left(u(0) - \int_0^x \frac{d}{dt} \left(e^{-\frac{1}{\epsilon} \int_0^t a(s)\, ds} \right) \frac{b(t)}{a(t)} dt \right) \\
&= e^{\frac{1}{\epsilon} \int_0^x a(s)\, ds} \left(u(0) - e^{-\frac{1}{\epsilon} \int_0^x a(s)\, ds} \frac{b(x)}{a(x)} + \frac{b(0)}{a(0)} \right. \\
&\quad \left. + \int_0^x e^{-\frac{1}{\epsilon} \int_0^t a(s)\, ds} \left(\frac{b(t)}{a(t)} \right)' dt \right) \\
&= -\frac{b(x)}{a(x)} + e^{\frac{1}{\epsilon} \int_0^x a(s)\, ds} \left(u(0) + \frac{b(0)}{a(0)} \right) + \int_0^x e^{\frac{1}{\epsilon} \int_t^x a(s)\, ds} \frac{d}{dt} \left(\frac{b(t)}{a(t)} \right) dt \\
&= -\frac{b(x)}{a(x)} - \frac{\epsilon}{a(x)} \frac{d}{dx} \left(\frac{b(x)}{a(x)} \right) \\
&\quad + e^{\frac{1}{\epsilon} \int_0^x a(s)\, ds} \left(u(0) + \frac{b(0)}{a(0)} + \frac{\epsilon}{a(0)} \frac{d}{dx} \left(\frac{b(x)}{a(x)} \right)_{x=0} \right) + O(\epsilon^2),
\end{aligned}
$$

we readily obtain the anticipated asymptotic approximation to any desired number of terms.

Next, applying the preceding result componentwise shows that the vector equation

$$
\epsilon v' = \wedge(x) v + k(x) \tag{3.315}
$$

has an asymptotic solution of the form

$$
v(x, \epsilon) = V(x, \epsilon) + e^{\frac{1}{\epsilon} \int_0^x \wedge(s)\, ds} (v(0) - V(0, \epsilon)) \tag{3.316}
$$

when the state matrix $\wedge$ is an $n \times n$ *diagonal* matrix with stable eigenvalues λ_i, and V is an outer expansion satisfying a system

$$
\epsilon V' = \wedge(x) V + k(x) \tag{3.317}
$$

as a regular power series in ϵ. Because $e^{\frac{1}{\epsilon} \int_0^x \wedge(s)\, ds}$ is diagonal with non-trivial decaying entries $e^{\frac{1}{\epsilon} \int_0^x \lambda_i(s)\, ds}$, the asymptotic solution of (3.315) is an additive function of the slow variable x and the fast variables $\frac{1}{\epsilon} \int_0^x \lambda_i(s)\, ds$, $i = 1, \ldots, n$.

More generally, consider the vector system

$$
\epsilon u' = A(x) u + b(x) \tag{3.318}
$$

when the state matrix A can be factored as

$$
A(x) = M(x) \wedge (x) M^{-1}(x) \tag{3.319}
$$

for a smooth invertible $n \times n$ matrix M and a diagonal matrix $\wedge$ with distinct smooth *stable* eigenvalues $\lambda_i(x)$, $i = 1, \ldots, n$. The *kinematic* change of variables

$$
u = M(x) v \tag{3.320}
$$

converts the equation (3.318) to the nearly diagonal form

$$
\epsilon v' = (\wedge - \epsilon M^{-1} M') v + M^{-1} b \tag{3.321}
$$

which has an asymptotic solution

$$v(x, \epsilon) = \mathsf{V}(x, \epsilon) + e^{\frac{1}{\epsilon} \int_0^x \wedge(s)\, ds}\, w(x, \epsilon) \qquad (3.322)$$

where $V(x, \epsilon)$ and $w(x, \epsilon)$ have power series expansions (cf. Lomov [298] and Wasow [513]). Their expansions can be obtained via undetermined coefficient methods.

If we can diagonalize the matrix $D(t)$, assuming it is stable, we can similarly treat the initial value problem for the slow-fast linear vector system

$$\begin{aligned} \dot{x} &= A(t)x + B(t)y \\ \epsilon\dot{y} &= C(t)x + D(t)y \end{aligned} \qquad (3.323)$$

and interpret the result in terms of using the slow time t and the fast times $\frac{1}{\epsilon} \int_0^t \lambda_i(s)\, ds$ where the $\lambda_i(t)$s are stable (nonrepeated) eigenvalues of $D(t)$. Block diagonalization of a conditionally stable matrix D might similarly allow us to treat certain two-point problems. Readers might look ahead to Example 16 in Chap. 6.

Historical Remarks

The work of Kaplun and Lagerstrom at Caltech in the 1950s was especially important to the development of matched expansions and its applications to fluid mechanics. Comparable work was simultaneously done by Proudman and Pearson at Cambridge University in England (cf. Proudman and Pearson [403]). Paco Lagerstrom (1914–1989), a Swedish-born Princeton math Ph.D., was Saul Kaplun (1924–1964)'s thesis advisor in 1954. (Contemporaries suggest they may have had an intimate personal, as well as professional, relationship.) The Polish-born Kaplun only published three papers, although Lagerstrom and others published a (not thoroughly edited) collection of his unfinished work as a monograph, Kaplun [237]. This and a book of reminiscences, *My Son Saul* [236] by his father and others and various memorials (at Caltech and Tel Aviv) made Kaplun into a hero of applied asymptotics in the 1960s. Lagerstrom persistently pursued their insights about matching for the next quarter century, using a limit process approach based upon the presumed overlapping domains of validity of the inner and outer expansions. Lagerstrom's book *Matched Asymptotic Expansions* [276] appeared in 1988, as he wrote,

> after a long sequence of earlier drafts.

Edward Fraenkel (in [155]) and Wiktor Eckhaus (in [133]) both insisted that existence of an overlap was not necessary for matching to succeed. Even after Lagerstrom's passing, Eckhaus [136] renewed the controversy, suggesting that the Kaplun extension theorem (intended to justify matching) could be based on Robinson's lemma in nonstandard analysis (cf. Diener and Diener [120]).

One fascinating example from Eckhaus [132], used in Lagerstrom [276], is the linear two-point problem

$$(\epsilon + x)u'' + u' = 1, \quad u(0) = 0, \ u(1) = 2. \tag{3.324}$$

Since x is an exact solution of the differential equation, one readily finds the exact solution

$$u(x, \epsilon) = x + \frac{\ln \epsilon - \ln(x + \epsilon)}{\ln \epsilon - \ln(1 + \epsilon)} \tag{3.325}$$

of the two-point problem. Note the initial layer.

Wiktor Eckhaus (1930–2000) was a significant contributor to both the nonlinear stability and singular perturbation literatures who had a number of productive students at Delft and Utrecht, including Ferdinand Verhulst, Johan Grasman, and Arjen Doelman, and the insightful early collaborator Eduardus de Jager. Verhulst, in turn, is known for his well-written texts on dynamical systems, averaging, singular perturbations and, most recently, Poincaré. He founded a publishing house, *Epsilon*, which produced mathematical monographs and textbooks in Dutch. Fortunately for most of us, later editions of many of its publications appeared in English from other publishers.

Chapter 4

Wendepunkts and Canards (Turning Points and Delayed Bifurcations)

(a) Simple Examples with Turning Points

We will soon realize that linear differential equations with turning points (Wendepunkts, in German) can feature complicated behavior. One of the best collections of illustrative examples (with sketches of the limiting solutions) is contained in the first chapter of a Dutch thesis, Hemker [202], which aimed to provide methods to solve such two-point singularly perturbed problems numerically. We will determine the limiting behavior of solutions to several examples using elementary methods.

1. Let's begin by considering

$$\epsilon y'' - x p(x) y' = 0 \quad \text{on } -1 \leq x \leq 1 \tag{4.1}$$

for a smooth function

$$p(x) > 0,$$

with bounded values $y(\pm 1)$ prescribed. Because $-xp(x) > 0$ for $x < 0$ and $-xp(x) < 0$ for $x > 0$, asymptotic matching might suggest a limiting solution of the form

$$y(x, \epsilon) \sim e^{-p(-1)(x+1)/\epsilon}(y(-1) - C) + C + e^{-p(1)(1-x)/\epsilon}(y(1) - C) \tag{4.2}$$

with $O(\epsilon)$-thick boundary layers at both ± 1 and with a constant outer limit C within $(-1, 1)$ (thereby satisfying the limiting equation $-xp(x)Y_0' = 0$). Nothing unusual then would happen at the simple turning point $x = 0$.

© Springer International Publishing Switzerland 2014
R. O'Malley, *Historical Developments in Singular Perturbations*,
DOI 10.1007/978-3-319-11924-3_4

The complication is that there is no obvious way using matching (rather than the exact solution) to specify the essential constant C. (One might even expect that the limit could change at $x = 0$, a singular point of the reduced equation.)

Since any nontrivial constant and

$$I(x) \equiv \int_0^x e^{\frac{1}{\epsilon} \int_0^t s\, p(s)\, ds}\, dt$$

are linearly independent solutions of the differential equation (4.1), however, we can directly seek a solution of the two-point problem as the linear combination

$$y(x, \epsilon) = c + I(x)d \tag{4.3}$$

for ϵ-dependent constants c and d. Then, the boundary conditions directly imply the linear equations

$$y(1) = c + I(1)d \text{ and } y(-1) = c + I(-1)d,$$

so Cramer's rule uniquely specifies

$$
\begin{cases}
c = \frac{I(1)y(-1) - I(-1)y(1)}{I(1) - I(-1)} \\
\text{and} \\
d = \frac{y(1) - y(-1)}{I(1) - I(-1)}.
\end{cases}
\tag{4.4}
$$

Within $(-1, 1)$, $\frac{I(x)}{I(1)-I(-1)}$ is asymptotically negligible, so the ϵ-dependent constant c provides the outer solution. The endpoint behavior depends on the limits of $\frac{I(\pm 1)}{I(1)-I(-1)}$. Under natural circumstances, we obtain the limit (4.2) suggested by matching, with the constant $C = c$. To determine c asymptotically, we need the limit of $\frac{I(-1)}{I(1)}$. The exact solution (4.3)–(4.4) is certainly preferable to the limiting asymptotic solution (4.2) since it, indeed, applies for all values of ϵ.

2. Hemker [202] also considered the two-point problem

$$
\begin{cases}
\epsilon y'' + 2xy' - 2y = 0, \quad -1 \le x \le 1 \\
\text{with} \\
y(-1) = -1 \text{ and } y(1) = 2.
\end{cases}
\tag{4.5}
$$

Since $2x < 0$ for $x < 0$, we might expect the limiting solution to be

$$y_L(x) = x \quad \text{for } x < 0,$$

since y_L is an exact solution of the differential equation and satisfies the left boundary condition. Likewise, the fact that $2x > 0$ for $x > 0$ satisfies the equation and the terminal condition suggests the limiting solution

$$y_R(x) = 2x \quad \text{for } x > 0.$$

Because both of these one-sided solutions are zero at $x = 0$ (they match!), we don't expect y to converge nonuniformly anywhere. However, $y'_L(x) = 1$ and $y'_R(x) = 2$ suggest that y' must jump at the turning point $x = 0$. (Some readers will realize that one could apply a Sturm transformation to eliminate the y' coefficient in (4.5) and then solve the resulting equation exactly using *parabolic cylinder* functions, as we shall do in the next section.)

The exact solution to (4.5) can also be found by reduction of order, since $y = x$ satisfies the differential equation. Its general solution will therefore have the form

$$y(x, \epsilon) = \alpha x + \beta \left(2x \int_{-1}^{x} e^{-s^2/\epsilon} \, ds + \epsilon e^{-x^2/\epsilon} \right) \tag{4.6}$$

for constants $\alpha(\epsilon)$ and $\beta(\epsilon)$. The boundary conditions then require that

$$\begin{cases} -1 = -\alpha + \epsilon \beta e^{-1/\epsilon} \\ \text{and} \\ 2 = \alpha + \beta \left(2 \int_{-1}^{1} e^{-s^2/\epsilon} \, ds + \epsilon e^{-1/\epsilon} \right). \end{cases}$$

Since $e^{-1/\epsilon}$ is asymptotically negligible, $\alpha \sim 1$ and $\beta \sim \frac{1}{2 \int_{-1}^{1} e^{-s^2/\epsilon} \, ds}$ provide the asymptotic solution

$$y(x, \epsilon) \sim x + \frac{2x \int_{-1}^{x} e^{-s^2/\epsilon} \, ds + \epsilon e^{-x^2/\epsilon}}{2 \int_{-1}^{1} e^{-s^2/\epsilon} \, ds}, \tag{4.7}$$

good up to an asymptotically negligible remainder. For $x < 0$, $e^{-x^2/\epsilon} \sim 0$ and

$$\frac{\int_{-1}^{x} e^{-s^2/\epsilon} \, ds}{\int_{-1}^{1} e^{-s^2/\epsilon} \, ds} \sim 0$$

since the primary contribution to the denominator comes from the vicinity of $s = 0$. Thus,

$$y \sim x \text{ for } x < 0,$$

as expected. For $x > 0$, however,

$$\frac{\int_{-1}^{x} e^{-s^2/\epsilon} \, ds}{\int_{-1}^{1} e^{-s^2/\epsilon} \, ds} \sim 1$$

shows that

$$y \sim 2x$$

there. Since

$$y'(x, \epsilon) \sim 1 + \frac{\int_{-1}^{x} e^{-s^2/\epsilon} \, ds}{\int_{-1}^{1} e^{-s^2/\epsilon} \, ds},$$

y' jumps from 1 to 2 in a $O(\sqrt{\epsilon})$-thick turning point (or shock layer) region about $x = 0$. Indeed, $y'(0, \epsilon) = 3/2$.

3. The equation

$$\epsilon y'' - xy' - y = 0 \tag{4.8}$$

has the general solution

$$y(x, \epsilon) = e^{\frac{x^2 - 1}{2\epsilon}} \left(A + B \int_0^x e^{-t^2/2\epsilon} dt \right). \tag{4.9}$$

(Check it.) If bounded values $y(\pm 1)$ are prescribed, we will need

$$y(1) = A + B \int_0^1 e^{-t^2/2\epsilon} dt$$

and

$$y(-1) = A - B \int_0^1 e^{-t^2/2\epsilon} dt.$$

This uniquely determines A and B and the exact solution

$$y(x, \epsilon) = \frac{1}{2} e^{\frac{x^2-1}{2\epsilon}} \left(y(1) + y(-1) + \left(\frac{\int_0^x e^{-t^2/2\epsilon} dt}{\int_0^1 e^{-t^2/2\epsilon} dt} \right) (y(1) - y(-1)) \right). \tag{4.10}$$

For $x > 0$, the ratio of integrals tends to 1, so the limiting solution is

$$y_R(x, \epsilon) \sim e^{\frac{x^2-1}{2\epsilon}} y(1) \tag{4.11}$$

there, i.e. there is an $O(\epsilon)$-thick terminal boundary layer near $x = 1$. For $x < 0$, the ratio of integrals tends to -1, so the limiting solution is

$$y_L(x, \epsilon) \sim e^{\frac{x^2-1}{2\epsilon}} y(-1), \tag{4.12}$$

providing an analogous initial boundary layer. Away from the endpoints, the outer solution satisfies

$$Y(x, \epsilon) = O(\epsilon^k) \text{ for all } k > 0.$$

Since the outer limit is trivial, we will later call the problem *nonresonant*.

4. Hemker [202] also considered the two-point Dirichlet problem for

$$\epsilon y'' + xy' - 2y = 0 \tag{4.13}$$

on $-1 \le x \le 1$. The sign of the coefficient x here suggests that there won't be endpoint layers. Since x^2 satisfies the reduced equation, the limiting solution should then be

$$x^2 y(1) \text{ for } x > 0 \quad \text{and} \quad x^2 y(-1) \text{ for } x < 0.$$

Because the resulting limits for y and y' match at $x = 0$, we must expect the limiting y'' to jump at the turning point when $y(1) \neq y(-1)$.

The equation (4.13) indeed has the exact even solution

$$y_1(x, \epsilon) \equiv x^2 + \epsilon$$

and the odd solution

$$y_2(x, \epsilon) \equiv (x^2 + \epsilon) \int_0^x e^{-s^2/2\epsilon} \, ds + \epsilon x e^{-x^2/2\epsilon},$$

so we write the general solution as the linear combination

$$y(x, \epsilon) = Ay_2(x, \epsilon) + By_1(x, \epsilon).$$

The boundary values $y(\pm 1)$ uniquely determine A and B and the exact solution

$$y(x, \epsilon) = \frac{1}{2} \frac{y_1(x, \epsilon)}{y_1(1, \epsilon)} (y(1) - y(-1)) + \frac{1}{2} \frac{y_2(x, \epsilon)}{y_2(1, \epsilon)} (y(1) + y(-1)). \qquad (4.14)$$

Since $\frac{y_1(x,\epsilon)}{y_1(1,\epsilon)} = \frac{x^2+\epsilon}{1+\epsilon} \to x^2$ as $\epsilon \to 0$, while

$$\frac{y_2(x, \epsilon)}{y_2(1, \epsilon)} = \frac{(x^2 + \epsilon) \int_0^x e^{-s^2/2\epsilon} \, ds + \epsilon x e^{-x^2/2\epsilon}}{(1 + \epsilon) \int_0^1 e^{-s^2/2\epsilon} \, ds + \epsilon e^{-1/\epsilon}} \to \begin{cases} x^2 & \text{for } x \geq 0 \\ -x^2 & \text{for } x \leq 0, \end{cases}$$

the limiting solution is that anticipated. Note the value of $y''(0)$ and that the interior shock layer for y'' at the turning point $x = 0$ has $O(\sqrt{\epsilon})$ thickness. Higher-order turning point problems are considered in Laforgue [274], Cheng [83], De Maesschalck [113], and Fruchard and Schäfke [165].

5. A much more challenging turning point problem was considered by Mahony and Shepherd [306] (cf. also O'Malley [371], which benefitted from the assistance of Grant Keady and Leonid Kalachev). They considered the initial value problem for the nonlinear equation

$$\epsilon y' = x^2(x^2 - y^2) \qquad (4.15)$$

on $x \geq -1$ with $y(-1) = 0$. Note that this Riccati equation can be converted to a linear second-order differential equation for $u = y'/y$, which can be solved in terms of *Bessel functions*. Alternatively, one can use MAPLE to find

$$y(x, \epsilon) = -x \left[\frac{K_{-\frac{5}{8}}\left(\frac{x^4}{4\epsilon}\right) - \lambda I_{-\frac{5}{8}}\left(\frac{x^4}{4\epsilon}\right)}{K_{\frac{3}{8}}\left(\frac{x^4}{4\epsilon}\right) + \lambda I_{\frac{3}{8}}\left(\frac{x^4}{4\epsilon}\right)} \right] \qquad (4.16)$$

for a constant λ (which might change at the turning point). The outer solutions

$$Y(x, \epsilon) \sim \begin{cases} -x - \frac{\epsilon}{2}\frac{1}{x^3} + \frac{7}{8}\frac{\epsilon^2}{x^7} + \dots & \text{for } -1 < x < 0 \\ x - \frac{\epsilon}{2}\frac{1}{x^3} + \dots & \text{for } 0 < x < 1 \end{cases} \qquad (4.17)$$

can be directly generated by iterating in the differential equation. (Note that the root $y = -x$ of the reduced equation $g(x, y) = x^4 - x^2 y^2 = 0$ is stable for $x < 0$ since $g_y(x, -x) = 2x^3 < 0$ there, while the root $y = x$ is stable for $x > 0$ since $g_y(x, x) = -2x^3 < 0$ there.) We can expect an $O(1)$ initial layer jump near $x = -1$ when $y(-1) \neq 1$. Behavior near $x = 0$ is complicated, due to the singularity of the outer limits (4.17) and because the solution (4.16) is a function of the stretched variable $x/\epsilon^{1/4}$, creating quite a thick corner layer at the turning point, as numerical computations confirm. The asymptotics of (4.15) are fascinating.

6. The two-point problem

$$\epsilon y'' + xy' - xy = 0, \quad 0 \leq x \leq 1, \quad y(0) = 0, \quad y(1) = e \qquad (4.18)$$

has an $O(\epsilon)$-thick initial layer and a complicated outer solution (cf. Bender and Orszag [36]). Fortunately, it can be solved exactly in terms of *parabolic cylinder functions*. One uses the *Sturm transformation*

$$y(x) = e^{-\frac{x^2}{4\epsilon}} u(t)$$

for $t = \frac{x+2\epsilon}{\sqrt{\epsilon}}$ to find u as a linear combination of

$$D_{-\epsilon}(it) \quad \text{and} \quad D_{-1-\epsilon}(-t). \qquad (4.19)$$

Using the usual asymptotic approximations (see Section (b)), one finds the outer limit

$$y(x, \epsilon) \sim \frac{e^x}{x^\epsilon} \qquad (4.20)$$

(as anticipated by Chen et al. [81]). Holzer and Kaper [211] point out that the matched expansion solution involves logarithmic switchback.

Exercise (Pearson [388])

(a) Determine the limiting behavior for

$$\epsilon y'' + xy' - \frac{1}{2}y = 0, \quad -1 \leq x \leq 1, \quad y(-1) = 1, \quad y(1) = 2.$$

Note that $y(0) = 0$ and that $y'(0)$ will be singular.

(b) Show, perhaps computationally, that the limiting solution to

$$\epsilon y'' + |x|y' - \frac{1}{2}y = 0, \quad y(-1) = 1, \quad y(1) = 2$$

is trivial for $-1 < x < 0$.

(b)   Boundary Layer Resonance

A prototypical *boundary layer resonance* problem consists of the linear equation

$$\epsilon y'' - xy' + \beta y = 0 \tag{4.21}$$

on $-1 \leq x \leq 1$, with prescribed boundary values $y(\pm 1)$ and a critical constant coefficient β. Note the turning point at $x = 0$ and that we have already exactly solved the problem when $\beta = 0$ and -1 (cf. (4.1) and (4.8)). Since the coefficient $-x$ is positive for $x < 0$ and negative for $x > 0$, we should expect endpoint layers at both ± 1. The limiting solution $Y_0(x)$ for $x \neq 0$ can be expected to satisfy

$$-xY_0' + \beta Y_0 = 0,$$

so

$$Y_0(x) = x^\beta C \tag{4.22}$$

for some constant C (possibly different for $x < 0$ and for $x > 0$). When $\beta > 0$, however, these outer limits match at the turning point. The ultimate boundary layer jumps at the endpoints $x = \pm 1$ will therefore be $y(1) - C$ and $y(-1) - (-1)^\beta C$, respectively, so if we suppose the same outer limits (4.22) for $x > 0$ and $x < 0$ and $O(\epsilon)$-thick boundary layers at both endpoints, the limiting solution (as obtained via matched expansions) will have the limiting composite form

$$y(x, \epsilon) = x^\beta C + e^{-(1+x)/\epsilon}(y(-1) - (-1)^\beta C) + e^{-(1-x)/\epsilon}(y(1) - C) + O(\epsilon) \tag{4.23}$$

uniformly in $-1 \leq x \leq 1$. An exact solution, in terms of special functions, will ultimately allow us to determine the unspecified constant C. A clever, but less straightforward, variational method (using (4.21) as an Euler–Lagrange equation) to find C was given by Grasman and Matkowsky [181], while a more subtle matching technique balancing asymptotically negligible terms was given in MacGillivray [302]. De Groen [185] observed that *resonance* (i.e., having a nontrivial limiting solution $x^\beta C$) is possible when $\beta(\epsilon)$ is asymptotically exponentially close to nonnegative integer *eigenvalues*. Of course, taking $C = 0$ when $\beta < 0$ avoids a singular Y_0.

The situation of having an ambiguous constant C in the representation (4.23) suggests a failure of the classical *matched* expansion technique, providing unusual interest to specialists anxious to retain confidence in matching.

The Sturm transformation

$$y = e^{x^2/4\epsilon} u \tag{4.24}$$

converts equation (4.21) to the *parabolic cylinder* (or Weber) *equation*

$$\epsilon u'' + \left(-\frac{x^2}{4\epsilon} + \frac{1}{2} + \beta \right) u = 0 \tag{4.25}$$

whose solutions can be expressed as a linear combination of the *parabolic cylinder functions* $D_\beta(x/\sqrt{\epsilon})$ and $D_\beta(-x/\sqrt{\epsilon})$ or $D_{-\beta-1}(ix/\sqrt{\epsilon})$ (to obtain a nonzero Wronskian and linear independence) (cf. Olver et al. [361]). These special functions are attributed to Weber and Hermite in White [519].

We will begin by setting

$$y(x,\epsilon) = e^{\frac{x^2}{4\epsilon}}\left(D_\beta\left(\frac{x}{\sqrt{\epsilon}}\right)A(\epsilon) + D_\beta\left(\frac{-x}{\sqrt{\epsilon}}\right)B(\epsilon)\right), \qquad (4.26)$$

where we will seek the constants A and B by imposing the prescribed boundary conditions. Whittaker and Watson [521] gives the asymptotic behavior of $D_\gamma(z)$ for $z \to \infty$ as

$$D_\gamma(z) \sim \begin{cases} z^\gamma e^{-z^2/4} & \text{for } |\arg z| < \frac{3\pi}{4} \\ z^\gamma e^{-z^2/4} - \frac{\sqrt{2\pi}}{\Gamma(-\gamma)}\frac{e^{i\pi\gamma}}{z^{\gamma+1}}e^{z^2/4} & \text{for } \frac{\pi}{4} < \arg z < \frac{5\pi}{4}. \end{cases}$$

(Note that these asymptotic limits change discontinuously across two so-called *Stokes lines* in the complex z plane. Figuring this out kept the Victorian Anglo-Irish Cambridge professor George Gabriel Stokes from sleep days prior to his wedding. As he quaintly wrote his fiancée:

> when the cat's away the mice can play. You are the cat and I am the poor little mouse. I have been doing what I guess You won't let me do when we are married, sitting up till 3 o'clock in the morning fighting hard against a mathematical difficulty. Some years ago I attacked an integral of Airy's, and after a severe trial reduced it to a readily calculable form. But there was a difficulty about it which, though I tried till I almost made myself ill, I could not get over and at last I have to give it up and profess myself unable to master it. I took it up again a few days ago, and after a two or three days' fight, the last of which I sat up till 3, I at last mastered it...

(cf. Ramis [405]).

A simple related example is provided by the function

$$f(\epsilon) = 1 + e^{-\frac{1}{\epsilon^2}} \qquad (4.27)$$

as $\epsilon \to 0$ in the complex plane. For $|\arg \epsilon| < \frac{\pi}{4}$ or $\frac{3\pi}{4} < \arg \epsilon < \frac{5\pi}{4}$, one has

$$f(\epsilon) \sim 1,$$

but for $\frac{\pi}{4} < \arg \epsilon < \frac{3\pi}{4}$ and $\frac{5\pi}{4} < \arg \epsilon < \frac{7\pi}{4}$,

$$f(\epsilon) \sim e^{-\frac{1}{\epsilon^2}}$$

becomes exponentially large. Thus the Stokes lines (called anti-Stokes lines by some) are given by $\arg \epsilon = \pm\frac{\pi}{4}$ and $\pm\frac{3\pi}{4}$. If you are reluctant to go complex, recall Hadamard's advice:

The shortest path between two truths of the real domain often passes through the complex one

(cf. Hadamard [191]). Also see Olde Daalhuis et al. [359]. Thus, the linear independence of $D_\beta(\pm\frac{x}{\sqrt\epsilon})$ in (4.26) is maintained as $\epsilon \to 0$ as long as $\Gamma(-\beta) \neq \infty$. When $\beta = N = 0, 1, 2, \ldots$, however,

$$e^{\frac{x^2}{4\epsilon}} D_N\left(\frac{x}{\sqrt\epsilon}\right) = He_N\left(\frac{x}{\sqrt\epsilon}\right),$$

the Nth *Hermite polynomial*, so we will then seek a solution of our two-point problem (4.21) in the alternative form

$$y(x, \epsilon) = He_N\left(\frac{x}{\sqrt\epsilon}\right) A(\epsilon) + e^{\frac{x^2}{4\epsilon}} D_{-N-1}\left(\frac{ix}{\sqrt\epsilon}\right) B(\epsilon) \qquad (4.28)$$

for new constants A and B.

For $\beta \neq N = 0, 1, 2, \ldots$, however, (4.26) implies the asymptotic limits

$$y(x, \epsilon) \sim \left(\frac{x}{\sqrt\epsilon}\right)^\beta A(\epsilon) + \frac{\sqrt{2\pi}}{\Gamma(-\beta)} \left(\frac{\sqrt\epsilon}{x}\right)^{\beta+1} e^{\frac{x^2}{2\epsilon}} B(\epsilon) \qquad \text{for } x > 0$$

and likewise for $x < 0$. Thus

$$(A, B) \sim \frac{\Gamma(-\beta)}{\sqrt{2\pi}} e^{-\frac{1}{2\epsilon}} \left(\frac{y(-1)}{(-\sqrt\epsilon)^{\beta+1}}, \frac{y(1)}{(\sqrt\epsilon)^{\beta+1}}\right)$$

in (4.26) and, thereby, the limits

$$y(x, \epsilon) \sim \begin{cases} \dfrac{e^{-(1-x^2)/2\epsilon}}{x^{\beta+1}} y(1) & \text{for } x > 0 \\[2mm] O\left(\dfrac{1}{\epsilon^{\frac{\beta+1}{2}}} e^{-\frac{1}{2\epsilon}}\right) & \text{for } x = 0 \\[2mm] \dfrac{e^{-(1-x^2)/2\epsilon}}{(-x)^{\beta+1}} y(-1) & \text{for } x < 0 \end{cases} \qquad (4.29)$$

featuring $O(\epsilon)$-thick endpoint boundary layers like $e^{-(1\mp x)/\epsilon} y(\pm 1)$ and a trivial outer limit (i.e., $C = 0$ in (4.22)) within $(-1, 1)$. Note that nothing unusual regarding y occurs at the turning point for these values of β.

For an integer $\beta = N \geq 0$, however, the representation (4.28) implies that

$$A(\epsilon) \sim \frac{y(1) + (-1)^N y(-1)}{2He_N(1/\sqrt\epsilon)}$$

and

$$B(\epsilon) \sim \frac{1}{2}\left(\frac{i}{\sqrt\epsilon}\right)^{N+1} e^{-\frac{1}{2\epsilon}} \left(y(1) - (-1)^N y(-1)\right),$$

yielding the uniform asymptotic limit

$$y(x, \epsilon) \sim \frac{1}{2} \left(y(1) + (-1)^N y(-1) \right) x^N$$
$$+ \frac{1}{2} \left(y(1) - (-1)^N y(-1) \right) \frac{e^{-(1-x^2)/2\epsilon}}{x^{N+1}}. \tag{4.30}$$

Note that the boundary layer behavior found here is more specific than, but consistent with, the behavior we would obtain by matching since, e.g., $e^{-(1-x^2)/2\epsilon} \sim e^{-(1-x)/\epsilon}$ near $x = 1$. Thus, we generally then have a nontrivial outer limit $x^N C$ for $|x| < 1$ with the specific constant

$$C = \frac{1}{2} \left(y(1) + (-1)^N y(-1) \right) \tag{4.31}$$

in (4.23). Again, nothing unusual happens to y at the turning point. As an example, the problem

$$\begin{cases} \epsilon y'' - xy' + y = 0, \\ y(-1) = -2, \quad y(1) = -1 \end{cases}$$

has the outer limit $x/2$ within $(-1, 1)$. The reason that we call this special case *resonance* is because we obtain a nontrivial limiting solution when $C \neq 0$ (cf. Ackerberg and O'Malley [3] and O'Malley [372]) for these special values of the parameter β. Note that Platte and Trefethen [392] use Chebychev approximations to successfully numerically solve the resonance problem for $\beta = 1$.

The equation

$$\epsilon y'' - xy' + \epsilon^n y = 0$$

will again be *nonresonant* for every integer $n > 0$ because $\Gamma(\epsilon^n)$ is only algebraically large. Indeed, for

$$\epsilon y'' - xy' + \beta(\epsilon)y = 0$$

when

$$\beta(0) = N = 0, 1, 2, \ldots,$$

resonance requires that *all* later coefficients β_j (for $j > 0$) in the asymptotic (Maclaurin) expansion of $\beta(\epsilon)$ be zero. On the other hand, we can also achieve resonance (i.e., a nonzero limiting solution) when $\beta(\epsilon) - N$ is appropriately asymptotically negligible. This is another example of *exponential asymptotics* or *asymptotics beyond all orders* (cf. Segur et al. [449]). Meanwhile, a direct calculation shows that the solution of the example

$$\begin{cases} \epsilon y'' + 2xy' + y = 0, \\ y(-1) = -1, \quad y(1) = 2 \end{cases}$$

becomes unbounded at the turning point $x = 0$.

The linear variable coefficient equation

$$\epsilon y'' - x\, h(x, \epsilon)y' + g(x, \epsilon)y = 0 \tag{4.32}$$

with

$$h(x, \epsilon) > 0$$

can be treated by a so-called *uniform reduction* method, a technique common in classical turning point theory (cf. Erdélyi [141]). We will first convert (4.32) to the form

$$\epsilon y_{\eta\eta} + F(\eta, \epsilon)y_\eta + G(\eta, \epsilon)y = 0 \tag{4.33}$$

by introducing the new independent variable

$$\eta = \sqrt{2 \int_0^x s\, h(s, 0)\, ds}. \tag{4.34}$$

Then, the Sturm transformation

$$z = e^{\frac{1}{2\epsilon} \int_0^\eta F(s,\epsilon)\, ds}\, y \tag{4.35}$$

implies that z will satisfy

$$\epsilon \frac{d^2 z}{d\eta^2} - \left(\frac{F^2}{4\epsilon} + \frac{1}{2}F_\eta - G \right) z = 0. \tag{4.36}$$

We then solve (4.36) asymptotically for $z(\eta, \epsilon)$ by finding power series for the coefficients $M(\eta, \epsilon)$, $N(\eta, \epsilon)$, and $\sigma(\epsilon)$ (termwise in ϵ) so that z is the linear combination

$$z = M(\eta, \epsilon)w + \epsilon N(\eta, \epsilon)w_\eta \tag{4.37}$$

where $w(\eta)$ satisfies the exactly solvable *comparison equation*

$$\epsilon \frac{d^2 w}{d\eta^2} - \left(\frac{\eta^2}{4\epsilon} + \sigma(\epsilon) \right) w = 0. \tag{4.38}$$

Resonance for (4.38) occurs precisely when $\sigma(\epsilon) = N$, a nonnegative integer (i.e., there are no higher-order corrections to σ as linear combinations of positive integer powers of ϵ). This initially provocative infinity of *necessary conditions for resonance* was labeled the *Matkowsky condition*, following Matkowsky [308]. The study was initiated by Ackerberg and O'Malley [3] and O'Malley [365], but was later continued by many prominent experts including W. Eckhaus, N. Kopell, H.-O. Kreiss, R. McKelvey, F. W. J. Olver, S. Parter, Y. Sibuya, and W. Wasow and their students and collaborators, among quite a few others worldwide. For more details regarding classical turning point theory, readers should consult Wasow [515] and Fedoryuk [145]. Sibuya [455, 456] provide more information about uniform reduction. De Groen [185] and [186] provides the eigenvalues and eigenfunctions for

$$\epsilon u'' - xh(x, \epsilon)u' + g(x, \epsilon)u = \lambda u, \quad u(\pm 1) = 0$$

and then sensibly uses *eigenfunction expansions* to study resonance. Related analysis is contained in Lee and Ward [284].

Instead of using parabolic cylinder functions, Holmes [209] directly uses *Kummer* functions. They satisfy the differential equation

$$y'' + \alpha xy' + \beta y = 0$$

and are the *confluent hypergeometric functions*

$$M\left(\frac{\beta}{2\alpha}, \frac{1}{2}, -\frac{\alpha}{2}x^2\right) \quad \text{and} \quad xM\left(\frac{\alpha+\beta}{2\alpha}, \frac{3}{2}, -\frac{\alpha}{2}x^2\right).$$

You might be led to these special functions by asking MAPLE to solve your differential equation. The *exit time problem* for stochastic processes is an important application with $\beta = 0$ (cf. Grasman and van Herwaarden [180] and Schuss [441]). For $\alpha = 1$, we can show that resonance provides unbounded solutions when $\beta = -2, -4, \ldots$ (cf. O'Malley [366]).

The general linear simple turning point problem

$$\epsilon y'' + f(x)y' + g(x)y = 0, \quad -1 \le x \le 1, \quad y(-1) = a, \quad y(1) = b \quad (4.39)$$

with

$$f(0) = 0 \quad \text{and} \quad f'(x) < 0 \quad (4.40)$$

was considered in Ackerberg and O'Malley [3] and de Jager and Jiang [224]. When

$$\beta \equiv -\frac{2g(0)}{f'(0)} \ne 2m, \quad m = 0, 1, 2, \ldots , \quad (4.41)$$

the outer limit is *trivial* and the boundary layer behavior is described by

$$y(x, \epsilon) \sim \begin{cases} ae^{-\frac{1}{\epsilon}\int_{-1}^{x} f(s)\, ds} & \text{for } x \text{ near } -1 \\ be^{-\frac{1}{\epsilon}\int_{1}^{x} f(s)\, ds} & \text{for } x \text{ near } 1. \end{cases} \quad (4.42)$$

In the exceptional case of *resonance*, whether there's one or two endpoint boundary layers depends on the sign of the integral

$$I = \int_{-1}^{1} f(s)\, ds. \quad (4.43)$$

If $I > 0$, the outer limit is given by

$$y(x, \epsilon) \sim bx^{\frac{\beta}{2}} e^{-\int_{x}^{1} \left[\frac{g(s)}{f(s)} + \frac{\beta}{2s}\right] ds} \quad \text{for } x > -1 \quad (4.44)$$

with the complementary initial layer. When $I < 0$, however,

$$y(x, \epsilon) \sim a(-x)^{\frac{\beta}{2}} e^{-\int_{-1}^{x} \left[\frac{g(s)}{f(s)} + \frac{\beta}{2s} \right] ds} \quad \text{for } x < 1 \qquad (4.45)$$

and for $I = 0$, as for (4.21), the outer limit is the average of (4.44) and (4.45). One shows this by constructing the WKB approximations

$$x^{\frac{\beta}{2}} e^{-\int^{x} \left[\frac{g(s)}{f(s)} + \frac{\beta}{2s} \right] ds} \qquad (4.46)$$

and

$$\frac{x^{-\frac{\beta}{2}}}{f(x)} e^{-\frac{1}{\epsilon} \int^{x} f(s) \, ds} e^{\int^{x} \left[\frac{g(s)}{f(s)} + \frac{\beta}{2s} \right] ds} \qquad (4.47)$$

(cf. Pearson [388]) and using turning point analysis or (like de Jager and Jiang) *two-timing* with the slow scale x and the fast scale $\frac{1}{\epsilon} \int^{x} f(s) \, ds$ (cf. Chap. 5).

(c) Canards

The following problems are ones for which classical Tikhonov–Levinson theory doesn't apply. The simplest example of a *canard* may be the initial value problem

$$\epsilon \dot{y} = ty, \quad t \geq -1, \quad y(-1) = 1. \qquad (4.48)$$

Here, the root $Y_0 = 0$ of the reduced equation

$$g(Y_0, t) \equiv tY_0 = 0$$

is stable for $t < 0$ (i.e., *attractive* since $g_y(y, t) = t < 0$), but unstable (i.e., *repulsive*) for $t > 0$ (since $g_y > 0$). The exact solution

$$y(t, \epsilon) = e^{(t^2 - 1)/2\epsilon} \qquad (4.49)$$

of (4.48), however, features an initial layer of $O(\epsilon)$ thickness near $t = -1$ and a trivial outer limit for $|t| < 1$. Moreover, it is symmetric about the turning point $t = 0$ and blows up exponentially after $t = 1$ for small ϵ. Such a *canard* (cf. Diener and Diener [120]) or *delayed bifurcation* occurs due to the asymptotically small size of the solution $e^{-\frac{1}{2\epsilon}}$ at $t = 0$. A canard is said to occur when the limiting trajectory sticks to the *repulsive* manifold for some time after a loss of stability. The term canard was (humorously) originally used because the phase-plane diagram for the van der Pol equation seemed to depict a duck (canard, in French) when wings were teasingly added (cf. Diener [122] and Chap. 5). By contrast to (4.48), the solution of the singularly perturbed equation

$$\epsilon \dot{y} = ty + \epsilon, \quad t \geq -1 \qquad (4.50)$$

with $y(-1) = 1$ blows up exactly at the turning point $t = 0$. Such a dramatic contrast between the solutions of (4.48) and (4.50) suggests that random perturbations might eliminate canards (cf. Berglund et al. [41]). The adventurous reader might check to see how various stiff integrators do at computing numerical solutions to (4.48) or (4.50). For a more sophisticated study of canards, see, for example, De Maesschalck [114]. An early reference is Benoit et al. [38].

The nonlinear example

$$\epsilon\dot{y} = ty^3, \quad t \geq -1, \quad y(-1) = 1 \tag{4.51}$$

has the exact solution

$$y(t, \epsilon) = \frac{1}{\sqrt{1 + \left(\frac{t^2-1}{2\epsilon}\right)}} \tag{4.52}$$

with the limiting solution $Y_0 = 0$ for $|t| < 1$. Gavin et al. [166] call this a *feeble* canard, because attraction and repulsion are now algebraic. Note that stiff integrators might do better here. Also note that Sobolev [468] is continuing a provocative study of canards and that Mishchenko at al. [323] gave Pontryagin credit for recognizing the phenomenon of a delay in the loss of stability, though other Russians suggest that this attribution represents loyalty to a thesis advisor (cf. Shishkova [453], however). Singularly perturbed logistic equations are considered in Verhulst [503]. An application to the FitzHugh–Nagumo equation is considered in Baer et al. [19].

Exercises

1. For the two-point problem

$$\begin{cases} \epsilon y'' + x\left(x^2 - \frac{1}{2}\right) y' - y = 0, \\ y(-1) = 1, \quad y(1) = 2, \end{cases}$$

   show that the limiting solution as $\epsilon \to 0^+$ is

$$Y_0(x) = \begin{cases} 1 - \frac{1}{2x^2}, & -1 \leq x \leq -\frac{\sqrt{2}}{2} \\ 0, & -\frac{\sqrt{2}}{2} \leq x \leq \frac{\sqrt{2}}{2} \\ 4 - \frac{2}{x}, & \frac{\sqrt{2}}{2} \leq x \leq 1. \end{cases}$$

2. For

$$\begin{cases} \epsilon y'' + \left(\frac{1}{2} - x^2\right) y' + xy = 0, \\ y(-1) = 1, \quad y(1) = 0, \end{cases}$$

   show (numerically or otherwise) that the limiting solution is

$$Y_0(x) = \begin{cases} \sqrt{2x^2 - 1}, & -1 \leq x \leq -\frac{\sqrt{2}}{2} \\ 0, & -\frac{\sqrt{2}}{2} \leq x \leq 1. \end{cases}$$

3. For

$$\begin{cases} \epsilon y'' + x^3 y' - y = 0 \\ y(-1) = 1, \quad y(1) = 2, \end{cases}$$

show that the limiting solution is

$$\begin{cases} e^{1-\frac{1}{x}}, & -1 \le x \le 0 \\ 2e^{1-\frac{1}{x}}, & 0 \le x \le 1 \end{cases}$$

(cf. Pearson [388] and Miranker [320] who seek numerical solutions to a variety of related problems.) A connection to Riccati transformations is explored in Dieci et al. [119].

4. (a) Solve the example

$$\epsilon y'' + xy' - y = 0, \quad y(0) \text{ and } y(1) \text{ prescribed}$$

exactly and determine the limiting solution.

(b) Solve the equation in terms of Kummer functions $M(a, b, z)$. Simplify your answer using the identities $M\left(-\frac{1}{2}, \frac{1}{2}, -\frac{1}{2}\alpha x^2\right) = e^{-\frac{\alpha}{2}x^2} + \alpha x \int_0^x e^{-\frac{\alpha}{2}r^2} dr$ and $M(0, b, z) = 1$.

Lin [292] considered the nonlinear two-point problem

$$\epsilon y'' - x(1 + y^2)y' - (1 + y^2)y = 0, \quad y(\pm 1) = 1 \tag{4.53}$$

numerically. The corresponding outer equation

$$xY_0' + Y_0 = 0$$

has nontrivial solutions $Y_0(x) = \frac{c}{x}$ that become unbounded at the turning point. We might seek an inner solution near $x = -1$ in the form

$$y^{in}(\sigma, \epsilon) \quad \text{for} \quad \sigma = \frac{1+x}{\epsilon}$$

such that $y^{in}(0) = 1$ and $y^{in}(\infty) = 0$ to match the trivial outer solution. The approximate local equation is

$$\frac{d^2 y^{in}}{d\sigma^2} + \left(1 + (y^{in})^2\right) \frac{dy^{in}}{d\sigma} = 0,$$

so integrating from infinity implies the Bernoulli equation $\frac{dy^{in}}{d\sigma} + y^{in} + \frac{1}{3}(y^{in})^3 = 0$. Thus,

$$y^{in}(\sigma) \sim \frac{3e^{-\sigma}}{\sqrt{4 - e^{-2\sigma}}}.$$

We will rewrite this as $y(x, \epsilon) \sim \frac{3e^{-(1+x)/\epsilon}}{\sqrt{4-e^{-2(1+x)/\epsilon}}}$. Doing likewise near $x = 1$ and combining the two asymptotic approximations, we obtain the uniformly valid approximation

$$y(x, \epsilon) \sim \frac{3e^{-(1-x^2)/2\epsilon}}{\sqrt{4 - e^{-(1-x^2)/\epsilon}}} \quad \text{for } -1 \le x \le 1. \tag{4.54}$$

Related examples are considered in Pearson [389].

The prominent Swedish numerical analyst Germund Dahlquist (1925–2005) introduced the solvable Riccati equation

$$\epsilon \dot{y} = ty - y^2, \quad t \ge -1, \quad y(-1) = 1 \tag{4.55}$$

as a challenging test problem for numerical integrators (cf. Dahlquist et al. [109]). The reduced equation

$$tY_0 - Y_0^2 = 0$$

has the root $Y_0(t) = 0$, stable for $t < 0$, and the root $Y_0(t) = t$, stable for $t > 0$, that cross at $t = 0$. The exact solution

$$y(t, \epsilon) = \frac{1}{e^{(1-t^2)/2\epsilon} + \frac{1}{\epsilon} \int_{-1}^{t} e^{(s^2 - t^2)/2\epsilon} ds} \tag{4.56}$$

of (4.55) can be evaluated asymptotically (using Laplace's method) to get the discontinuous limiting solution

$$y(t, \epsilon) \sim \begin{cases} 0, & -1 < t < 1 \\ \frac{1}{3}, & t = 1 \\ t, & t > 1 \end{cases} \tag{4.57}$$

(known as *Dahlquist's knee*) with a boundary layer at $t = -1$ and a shock layer at $t = 1$. Note, again, the occurrence of a canard or *delayed bifurcation* for $0 < t < 1$.

The example

$$\epsilon \dot{y} = t(1 - y^2), \quad t \ge -1, \quad y(-1) = \frac{1}{2} \tag{4.58}$$

was introduced by Mahony and Shepherd [306]. The separable equation has the exact solution

$$y(t, \epsilon) = \frac{3e^{\frac{t^2-1}{\epsilon}} - 1}{3e^{\frac{t^2-1}{\epsilon}} + 1}. \tag{4.59}$$

When $t^2 < 1$, the limiting solution is $Y_0(t) = -1$, but for $t^2 > 1$, it is $Y_0(t) = 1$. If we instead had $y(-1) < -1$, we'd again get the limit -1 for $|t| < 1$, but blowup would occur for $|t| > 1$. Lebowitz and Schaar [283] considered the related *exchange of stability* in systems without bifurcation delay. Also see Nefedov and Schneider [346].

Verhulst [500] considers the separable (Riccati) equation

$$\epsilon\dot{y} = -\sin ty(1-y), \quad y(0) = 1/2. \tag{4.60}$$

It has the exact 2π-periodic solution

$$y(t, \epsilon) = \frac{e^{-\frac{1}{\epsilon}(1-\cos t)}}{1 + e^{-\frac{1}{\epsilon}(1-\cos t)}}. \tag{4.61}$$

Since the exponentials become negligible away from $t = 2\pi k$, the solution tends to the trivial root of the reduced equation, except near $t = 2\pi k$ for integers k, where it spikes to the value $1/2$. The root $Y_0 = 1$ of the reduced equation is never reached for this initial value.

We note that other interesting examples are given by Fruchard and Schäfke [165]. They provide necessary and sufficient conditions for resonance for a broader variety of equations and they give composite expansions for their asymptotic solutions.

Kreiss and Parter [264] provide many good examples, including the readily solvable equations

$$\epsilon y'' - 2xy' = 0, \quad -1 \le x \le \frac{1}{4}$$

$$\epsilon y'' - \frac{1}{2}y + b_0(\epsilon)y = 0, \quad \frac{1}{4} \le x \le \frac{1}{2}$$

with $b_0(\epsilon)$ exponentially small.

Among many intriguing examples found in the literature, one of the most interesting involves *small exponent asymptotics* (cf. Fowler et al. [154] and Kember et al. [246]). They consider the solvable initial value problem

$$\dot{y} = \frac{1}{y^\epsilon} - 2, \quad y(-1) = 1 \tag{4.62}$$

for ϵ small and encounter a "hidden" boundary layer at $t = 0$. An implicit solution $t(y)$ in straightforward.

Exercises

1. Determine the limiting behavior for Dahlquist's example

$$\epsilon\dot{y} = (1-t)y - y^2, \quad t \ge 0, \quad y(0) = \frac{1}{2}.$$

2. For the problem

$$\epsilon\dot{y} = ty - y^2, \quad t \ge -2,$$

show that

$$y(t, \epsilon) \sim \begin{cases} 0, & -2 < t < 2 \\ t - \frac{\epsilon}{t} + \dots, & t > 2. \end{cases}$$

Note that these trivial and nontrivial *slow* manifolds can be generated termwise by substituting a regular perturbation series

$$Y(t, \epsilon) = Y_0(t) + \epsilon Y_1(t) + \dots$$

into the differential equation.

3. Dahlquist [108] considers the Riccati equation

$$\epsilon \dot{y} = \sin^2 t - y^2, \quad y(0) > 0.$$

Use a matching argument near $t = \pi$ to show that the asymptotic solution satisfies $y(t, \epsilon) = |\sin t| + O(\sqrt{\epsilon})$, $t > 0$.

4. Lin and O'Malley [293] show that the solution to the separable equation

$$\epsilon \dot{y} = \frac{ty}{1+y}, \quad y(-1) = 1$$

is given by

$$ye^{y-1} = e^{-\frac{1}{2\epsilon}(1-t^2)}.$$

Show this and verify that the limiting solution is trivial for $|t| < 1$. (Some might express the solution in terms of the *Lambert W function*.)

5. Consider the two-point problem

$$\epsilon y'' - xy' + (x - \epsilon)y = 0, \quad 0 \le x \le 1$$

with $y(0)$ and $y(1)$ prescribed.

(a) Noting that e^x is an exact solution of the differential equation, find the exact solution of the boundary value problem.

(b) Noting that $e^x y(0)$ is the outer solution, determine the nature of the boundary layer behavior near $x = 1$.

Chapter 5

Two-Timing, Geometric, and Multi-scale Methods

(a) Elementary Two-Timing

The Brooklyn native Julian Cole (1925–1999) got his Ph.D. in aeronautics at Caltech in 1949 with Hans Liepmann (a German émigre of 1939) as his advisor. He remained on the Caltech faculty until 1968 where he maintained active contact with Kaplun, Lagerstrom and others in aeronautics, applied mathematics and industry, attempting to understand singular perturbations more deeply and to apply its growing methodology. Then he moved to UCLA and, ultimately, Rensselaer. He and his Jerusalem-born student Jerry Kevorkian, who spent his academic career at the University of Washington, developed and applied asymptotic methods involving two- (i.e. multi-) time or multiple scales in the early 1960s (cf. the obituary of Cole by Bluman et al. [48]). Related approaches were made by the Soviet Kuzmak [273], the Australian Mahony [305], and the American Cochran [89], among others, but Cole and Kevorkian had the dominant long-term impact. The previously cited work of Lomov is also recommended reading, as is the paper by Levey and Mahony [286]. The monograph *Perturbation Methods in Applied Mathematics*, Cole [92], considered singular perturbations in a broad applied math setting, where both the development of the underlying techniques and significant and diverse applications were included. The book approaches matching using intermediate limits and presumes a corresponding overlap of inner and outer domains. The examples used are generally very instructive and quite nontrivial.

Two-variable expansions are naturally introduced, for example, in the presence of a small disturbance acting cumulatively for a long time. The simultaneous occurrence of a small parameter and a long time interval implies

© Springer International Publishing Switzerland 2014   141
R. O'Malley, *Historical Developments in Singular Perturbations*,
DOI 10.1007/978-3-319-11924-3_5

that we, indeed, encounter a *two-parameter* singular perturbation *problem* (cf. O'Malley [366]). Smith [466] aptly called such initial value problems singular perturbations with a nonuniformity at infinity. The two-timing approach generalizes and extends the classical Poincaré–Lindstedt method of *strained coordinates* (cf., e.g., Poincaré [394], Minorsky [319], and Murdock [335]) which is often applied to solve Duffing's equation and to describe related nonlinear oscillations. The procedure ultimately provides the same asymptotic expansion for the solution as the method of averaging (cf. Bogoliubov and Mitropolsky [51]), which is long known to be mathematically justified. Greatly expanded editions of Cole's book, coauthored with Kevorkian, appeared in 1981 and 1996 [247, 248]. The late Peter Chapman, a colleague of Mahony in Perth, developed a promising manuscript [77] that may never have been finished. It explained the ongoing work of Mahony and the paper of Kuzmak. Mahony's work, more generally, is reviewed in Fowkes and Silberstein [153]. The major recent generalization from two-timing to multiscale modeling is considered in E [130].

As a first example, we shall describe the application of two-timing to the nearly linear *Rayleigh equation*

$$\ddot{y} + y = \epsilon\left(\dot{y} - \frac{1}{3}\dot{y}^3\right) \quad \text{on } t \geq 0 \tag{5.1}$$

with initial values

$$y(0) = 0 \quad \text{and} \quad \dot{y}(0) = 1. \tag{5.2}$$

This, presumably, describes the oscillations of a clarinet reed (cf. Rayleigh [409]). Note that if y satisfies the Rayleigh equation, $\dot{y}$ will satisfy the van der Pol equation (which we will later study).

Two-timing anticipates that the solution of (5.1–5.2) will evolve, depending on both the given fast-time t and the introduced *slow time*

$$\tau \equiv \epsilon t \tag{5.3}$$

using a formal *two-time* power series *expansion*

$$y(t, \epsilon) = Y(t, \tau, \epsilon) \sim Y_0(t, \tau) + \epsilon Y_1(t, \tau) + \epsilon^2 Y_2(t, \tau) + \dots \tag{5.4}$$

when $t = O(1/\epsilon)$, i.e. $\tau = O(1)$. (Don't be confused because we used $\tau = \frac{t}{\epsilon}$ as a fast time in Chap. 3.) Sophisticates will realize that (5.4) is a generalized asymptotic expansion, since its coefficients Y_k depend on ϵ through τ. The chain rule requires that

$$\dot{y} = Y_t + \epsilon Y_\tau \quad \text{and} \quad \ddot{y} = Y_{tt} + 2\epsilon Y_{t\tau} + \epsilon^2 Y_{\tau\tau},$$

so equation (5.1) implies that the two-time expansion (5.4) must satisfy the partial differential equation

$$Y_{tt} + Y + \epsilon\left(2Y_{t\tau} - Y_t + \frac{1}{3}Y_t^3\right) + \epsilon^2(Y_{\tau\tau} - Y_\tau + Y_t^2 Y_\tau) + \epsilon^3 Y_t Y_\tau^2 + \frac{\epsilon^4}{3}Y_\tau^3 = 0. \tag{5.5}$$

Converting the ODE (5.1) to the PDE (5.5) may not, at first, seem like a step forward, but wait and experience its success. Equating coefficients of successive powers of ϵ as a regular perturbation expansion requires

$$Y_{0tt} + Y_0 = 0, \tag{5.6}$$

$$Y_{1tt} + Y_1 = -2Y_{0t\tau} + Y_{0t} - \frac{1}{3}Y_{0t}^3, \tag{5.7}$$

etc. From (5.6), it follows that Y_0 must be a linear combination of $\cos t$ and $\sin t$ with τ-dependent coefficients, so we set

$$Y_0(t, \tau) = A_0(\tau) \cos t + B_0(\tau) \sin t, \tag{5.8}$$

where A_0 and B_0 so far remain undetermined, except for their initial values since

$$\begin{cases} y(0) = Y_0(0,0) = A_0(0) = 0 \\ \text{and} \\ \dot{y}(0) \sim Y_{0t}(0,0) = B_0(0) = 1. \end{cases} \tag{5.9}$$

Thus, the representation (5.8) introduces *amplitudes* A_0 and B_0 that are slowly varying functions of t.

Next, using the partial differential equation (5.7) for Y_1, we require

$$\begin{aligned} Y_{1tt} + Y_1 = &- 2\left(-\frac{dA_0}{d\tau} \sin t + \frac{dB_0}{d\tau} \cos t\right) \\ &+ (-A_0 \sin t + B_0 \cos t) \\ &- \frac{1}{3}(-A_0 \sin t + B_0 \cos t)^3. \end{aligned} \tag{5.10}$$

Recalling the trigonometric identities $\sin^3 t = \frac{3}{4} \sin t - \frac{1}{4} \sin 3t$, $\sin^2 t \cos t = \frac{1}{4} \cos t - \frac{1}{4} \cos 3t$, $\sin t \cos^2 t = \frac{1}{4} \sin t + \frac{1}{4} \sin 3t$, and $\cos^3 t = \frac{3}{4} \cos t + \frac{1}{4} \cos 3t$, we find that $Y_{1tt} + Y_1$ is a linear combination of $\sin t$, $\cos t$, $\sin 3t$, and $\cos 3t$, with τ-dependent coefficients. The general solution for Y_1 follows simply by the method of undetermined coefficients. Multiples of $\sin t$ and $\cos t$ in the forcing term yield unbounded responses like $t \sin t$ and $t \cos t$ in the particular solution. Since $\sin t$ and $\cos t$ are solutions of the homogeneous equation, the presence of such terms in the forcing is said to *resonate* with the complementary solutions for Y_1. Such *secular* response terms can't be allowed if the expansion $Y(t, \tau, \epsilon)$ is to remain asymptotic for large t. Specifically, since (5.10) implies that

$$\begin{aligned} Y_{1tt} + Y_1 = &\left(2\frac{dA_0}{d\tau} - A_0 + \frac{A_0^3}{4} + \frac{A_0 B_0^2}{4}\right) \sin t \\ &+ \left(-2\frac{dB_0}{d\tau} + B_0 - \frac{A_0^2 B_0}{4} - \frac{B_0^3}{4}\right) \cos t \\ &+ \left(-\frac{A_0^3}{12} + \frac{A_0 B_0^2}{4}\right) \sin 3t + \left(\frac{A_0^2 B_0}{4} - \frac{B_0^3}{12}\right) \cos 3t, \end{aligned} \tag{5.11}$$

to make the first harmonics disappear in the forcing thereby requires A_0 and B_0 to satisfy the coupled vector initial value problem

$$\begin{cases} 2\frac{dA_0}{d\tau} = A_0 - \frac{A_0}{4}(A_0^2 + B_0^2), & A_0(0) = 0 \\ \text{and} \\ 2\frac{dB_0}{d\tau} = B_0 - \frac{B_0}{4}(A_0^2 + B_0^2), & B_0(0) = 1. \end{cases} \tag{5.12}$$

Uniqueness implies that

$$A_0(\tau) = 0 \tag{5.13}$$

while the explicit solution of the remaining Bernoulli equation determines

$$B_0(\tau) = \frac{2}{\sqrt{1 + 3e^{-\tau}}}, \tag{5.14}$$

which remains defined for all $\tau \geq 0$.

Thus, Y_1 must be a bounded solution of

$$Y_{1tt} + Y_1 = -\frac{B_0^3}{12} \cos 3t.$$

A particular solution as a slowly varying multiple of $\cos 3t$ follows using undetermined coefficients and since its complementary solution will be a linear combination of $\cos t$ and $\sin t$, Y_1 will have the form

$$Y_1(t, \tau) = A_1(\tau) \cos t + B_1(\tau) \sin t + \frac{B_0^3(\tau)}{96} \cos 3t.$$

Moreover, A_1 and B_1 must satisfy the initial conditions

$$0 = Y_1(0, 0) = A_1(0) + \frac{B_0^3(0)}{96} \quad \text{and} \quad 0 = Y_{1t}(0, 0) + Y_{0\tau}(0, 0) = B_1(0) + A_0(0),$$

so

$$A_1(0) = -\frac{1}{96} \quad \text{and} \quad B_1(0) = 0.$$

We will completely specify A_1 and B_1 as an exercise. Equations (5.13–5.14) determines the limiting two-time approximation

$$y(t, \epsilon) = Y_0(t, \tau) + O(\epsilon) \tag{5.15}$$

on $0 \leq \tau < \infty$ with

$$Y_0(t, \tau) = \frac{2 \sin t}{\sqrt{1 + 3e^{-\tau}}}. \tag{5.16}$$

Proofs justifying the two-time technique by proving the estimate (5.15) are given, e.g., in Smith [466]. They can be expected to hold with some tradeoffs both on longer time intervals and to higher-order. Note that Poincaré in the preface to the first volume of *Celestial Mechanics* [394] reported

> All efforts of geometers in the second half of this century have had as main objective the elimination of secular terms.

Exercise

Knowing Y_0, determine Y_1 completely by eliminating resonant terms in the differential equation

$$Y_{2tt} + Y_2 + (2Y_{1t\tau} - Y_1 + Y_{0t}^3 Y_{1t}) + Y_{0\tau\tau} - Y_{0\tau} + Y_{0t}^2 Y_{0\tau} = 0$$

for Y_2 by appropriately determining the first harmonic coefficients $A_1(\tau)$ and $B_1(\tau)$ of Y_1.

Paul Germain [169] introduces *multiple scales* more broadly:

> when a physical phenomenon is thought to be represented by the occurrence of steep gradients in one variable only .... Assume that the manifold across which the gradients are steep is $F(x, t) =$ constant. Then, the mathematical progressive wave structure is
>
> $$U(t, x, \frac{F}{\epsilon}, \epsilon)$$
>
> with $\xi \equiv \frac{F}{\epsilon}$ considered as a fifth variable.

See Germain [169] and Zeytounian [533] for more details.

Historical Comment

Poincaré won the (Swedish and Norwegian) *King Oscar II Prize* in 1889 (celebrating the king's 60th birthday) for his work on the three-body problem. The hastily prepared submission was, however, in error (cf. Barrow-Green [29]). (Distributed copies were collected and trashed, but Barrow-Green recently found one remaining in the Mittag-Leffler Institute library. Mittag-Leffler was a judge (together with Hermite and Weierstrass), organizer of the prize, and founding editor of *Acta Mathematica*.) A corrected version of Poincaré's paper was published in *Acta Mathematica* in 1890 (at Poincaré's expense). Much of the difference relates to whether the formal solutions found by the astronomer Lindstedt in 1883 were convergent or simply asymptotic (and to the nonintegrability of the three-body problem). This distinction was further highlighted in the introduction to the second volume of Poincaré's *Celestial Mechanics* [394] where a difference between "astronomers" and "mathematicians" may be understood if we realize that astronomers traditionally call asymptotic series convergent. Many years later, KAM theory (the research of Kolmogorov, Arnold, and Moser from 1954 to 1963 on the persistence of quasi-periodic motions under small perturbations) shows that some similar results actually converge (cf. Arnold et al. [12]). Curiously, Szpiro [478] suggests that the Swedes Lindstedt and Gyldén each claimed some of their work had precedence over Poincaré's, but this is not noted by the more scholarly Barrow-Green. More details can be found in Charpentier et al. [78], Verhulst [502], and Gray [182].

According to Stubhaug [477], Gyldén, head of the Stockholm Observatory, characterized the entire prize as

humbug.

Ziegler [537] reports that Mittag-Leffler tried to get a Nobel prize for Poincaré from their initiation in 1901 until Poincaré's death. Stubhaug emphasizes that Mittag-Leffler wanted to promote theoretical physics for the prize as well as Poincaré. (It's never been clear why there's no mathematics Nobel, but there's now an *Abel prize* and other, even bigger, new ones.)

Now consider the initial value problem

$$\ddot{y} + y + \epsilon y^3 = 0, \quad y(0) = 1, \quad \dot{y}(0) = 0 \tag{5.17}$$

for *Duffing's equation* on $t \geq 0$ (Kovacic and Brennan [261] provides a brief biography of Georg Duffing (1861–1944) as well as a partial translation of his 1918 book *Forced Oscillations with Variable Natural Frequency and their Technical Significance*, originally published by Sammlung Vieweg, Braunschweig, in German.). Using a phase-plane analysis, for example, one can readily convince oneself that the solution is periodic (cf. Mudavanhu et al. [332]). A regular perturbation expansion produces artificial secular terms that are clearly spurious, so we might naturally instead seek an asymptotic solution

$$y(t, \epsilon) = z(s, \epsilon) \sim z_0(s) + \epsilon z_1(s) + \ldots \tag{5.18}$$

as a function of a so-called *strained* coordinate

$$s = (1 + \epsilon \Omega(\epsilon))t \tag{5.19}$$

where the frequency perturbation

$$\Omega(\epsilon) \sim \Omega_0 + \epsilon \Omega_1 + \ldots \tag{5.20}$$

for constants Ω_j is to be determined termwise to achieve periodicity of the terms of the expansion for z with respect to the strained s. Clearly, z will need to satisfy the initial value problem

$$(1 + \epsilon \Omega(\epsilon))^2 \frac{d^2 z}{ds^2} + z + \epsilon z^3 = 0, \quad z(0, \epsilon) = 1, \quad \frac{dz}{ds}(0, \epsilon) = 0. \tag{5.21}$$

Proceeding termwise, we obtain successive initial value problems

$$\frac{d^2 z_0}{ds^2} + z_0 = 0, \quad z_0(0) = 1, \quad \frac{dz_0}{ds}(0) = 0,$$

$$\frac{d^2 z_1}{ds^2} + z_1 + 2\Omega_0 \frac{d^2 z_0}{ds^2} + z_0^3 = 0, \quad z_1(0) = 0, \quad \frac{dz_1}{ds}(0) = 0,$$

etc. Since
$$z_0(s) = \cos s, \tag{5.22}$$

z_1 must satisfy
$$\frac{d^2 z_1}{ds^2} + z_1 = \left(2\Omega_0 - \frac{3}{4}\right)\cos s - \frac{1}{4}\cos 3s.$$

To avoid secular terms in z_1, we must pick
$$\Omega_0 = 3/8 \tag{5.23}$$

to obtain
$$z_1(s) = -\frac{1}{32}\cos s + \frac{1}{32}\cos 3s. \tag{5.24}$$

At the next stage, we find
$$s = \left(1 + \frac{3\epsilon}{8} - \frac{21}{256}\epsilon^2 + \dots\right)t \tag{5.25}$$

and
$$y(t, \epsilon) = z(s, \epsilon) = \left(1 - \frac{\epsilon}{32} + \frac{23}{1024}\epsilon^2 + \dots\right)\cos s$$
$$+ \epsilon\left(\frac{1}{32} - \frac{\epsilon}{64} + \dots\right)\cos 3s \tag{5.26}$$
$$+ \epsilon^2\left(\frac{1}{1024} + \dots\right)\cos 5s + \dots.$$

The series for the frequency actually converges. Andersen and Geer [7] calculated series for the corresponding periodic solution of the van der Pol equation to $O(\epsilon^{24})$ terms using MACSYMA and to $O(\epsilon^{164})$ terms via Taylor series.

Note that use of the coordinate s (which is only ever approximated as a polynomial of increasing order in ϵ) corresponds to a *multitime* expansion using the times t, ϵt, $\epsilon^2 t, \dots$. (See below.) Later scales are not needed when we bound $\epsilon^k t$ for any fixed k.

More generally, nonlinear *clock functions*
$$\tau_i(t, \epsilon), \quad i = 0, 1, \dots, N$$

are sometimes used to determine multitime expansions
$$x(t, \epsilon) = X(\tau_0, \tau_1, \dots, \tau_N, \epsilon).$$

Ablowitz [1] points out that this *frequency-shift* method, that he calls the Stokes-Poincaré approach, is limited to equations in conservation form (see Sect. (c)). (It does not apply, e.g., to the van der Pol equation, though it will provide its periodic limit cycle.)

More generally, when one considers the nearly linear equation

$$\ddot{y} + y + \epsilon f(y, \dot{y}) = 0, \tag{5.27}$$

two-timing produces a first term approximation

$$Y_0(t, \tau) = A_0(\tau) \cos t + B_0(\tau) \sin t \tag{5.28}$$

for $\tau = \epsilon t$ and requires the second term to satisfy a resulting nonhomogeneous equation

$$\frac{\partial^2 Y_1}{\partial t^2} + Y_1 = P(t, \tau) \tag{5.29}$$

determined in terms of A_0 and B_0. The *Fredholm alternative* requires the two orthogonality conditions,

$$\int_0^{2\pi} P(t, \tau) \cos t \, dt = 0 \quad \text{and} \quad \int_0^{2\pi} P(t, \tau) \sin t \, dt = 0, \tag{5.30}$$

which coincide with the differential equations for A_0 and B_0 needed to eliminate secular terms in Y_1. An alternative representation

$$Y_0(t, \tau) = C_0(\tau) \cos(t + D_0(\tau))$$

with slowly varying amplitude C_0 and phase D_0 (instead of (5.28)) would also be effective. Kevorkian and Cole [249] use multiple scale methods for a variety of problems.

Exercise (cf. Hinch [206])

Show that Duffing's equation could also be solved by directly seeking a solution of the form

$$y(t, \epsilon) = z(s, \epsilon) \tag{5.31}$$

where s is determined by inverting a *near-identity transformation*

$$t = s + \epsilon t_1(s) + \epsilon^2 t_2(s) + \dots \tag{5.32}$$

for functions $t_j(s)$ that provide a periodic solution $z(s, \epsilon)$ termwise.

One can obtain Mathieu's equation

$$\frac{d^2 x}{dt^2} + (\delta + \epsilon \cos t)x = 0 \tag{5.33}$$

as a linearization of Duffing's equation (cf. Jordan and Smith [230]). Moreover, one can study what parameter values δ and ϵ provide bounded or unbounded solutions. Transitions of solutions $x(t, \epsilon)$ from stability to instability occur along curves $\delta(\epsilon)$ called *tongues* that can be obtained using perturbation methods (cf. Nayfeh and Mook [345]). Bifurcations may involve *hidden* time scales, like $\epsilon^{3/2}t$ (cf. Chen et al. [81] and Verhulst [504]). The traditional two-time expansion

$$Y(t,\tau,\epsilon) \sim Y_0(t,\tau) + \epsilon Y_1(t,\tau) + \ldots \tag{5.34}$$

is effective for equation (5.27). We can expect such results to hold for bounded τ values, though Greenlee and Snow [184] showed that with appropriate damping, two-timing is valid on the whole half-line $t \geq 0$.

We can also apply multiple scales to equations with boundary layer behavior, even though Lagerstrom [276] sought a dichotomous distinction between *layer type* and *secular* problems. Consider, for example, the nonlinear two-point boundary value problem

$$\epsilon y'' + a(x)y' + g(x,y) = 0 \tag{5.35}$$

on $0 \leq x \leq 1$ where

$$a(x) > 0, \tag{5.36}$$

a and g are smooth, and bounded end values

$$y(0) \quad \text{and} \quad y(1) \tag{5.37}$$

are prescribed. We anticipate having an initial boundary layer due to the sign of a, so we introduce the *stretched* (fast) *variable*

$$\eta = \frac{1}{\epsilon} \int_0^x a(s)\, ds \tag{5.38}$$

(known to be appropriate for the corresponding linear equations) and seek an asymptotic solution to the two-point problem for (5.35) in the two-variable (or multiscale) form

$$y(x,\epsilon) = Y(x,\eta,\epsilon) \sim Y_0(x,\eta) + \epsilon Y_1(x,\eta) + \ldots. \tag{5.39}$$

Since

$$y' = Y_x + \frac{a(x)}{\epsilon} Y_\eta$$

and

$$\epsilon y'' = \epsilon Y_{xx} + 2a(x)Y_{x\eta} + a'(x)Y_\eta + \frac{a^2(x)}{\epsilon} Y_{\eta\eta},$$

the ordinary differential equation (5.35) for y requires Y to satisfy the partial differential equation

$$\frac{a^2(x)}{\epsilon}(Y_{\eta\eta} + Y_\eta) + (2a(x)Y_{x\eta} + a'(x)Y_\eta)$$
$$+ a(x)Y_x + g(x,Y)) + \epsilon Y_{xx} = 0 \tag{5.40}$$

as a regular perturbation series, with the coefficients Y_k in (5.39) depending on x and η simultaneously. Equating successive coefficients in (5.40) to zero then requires that

$$Y_{0\eta\eta} + Y_{0\eta} = 0, \tag{5.41}$$

$$Y_{1\eta\eta} + Y_{1\eta} + \frac{1}{a^2(x)}\left(2a(x)Y_{0x\eta} + a'(x)Y_{0\eta} + a(x)Y_{0x} + g(x, Y_0)\right) = 0, \tag{5.42}$$

etc. Equation (5.41) implies that Y_0 is a linear combination of 1 and $e^{-\eta}$, with undetermined coefficients depending on the slow variable x. Thus, we set

$$Y_0(x, \eta) = A_0(x) + B_0(x)e^{-\eta}. \tag{5.43}$$

The boundary values then require that

$$A_0(0) + B_0(0) = y(0) \quad \text{and} \quad A_0(1) \sim y(1) \tag{5.44}$$

since $e^{-\eta}$ is asymptotically negligible outside the initial layer. The equation (5.42) then requires that

$$\begin{aligned}
Y_{1\eta\eta} + Y_{1\eta} + \frac{1}{a^2(x)} &\left(-2a(x)B_0'(x)e^{-\eta} - a'(x)B_0(x)e^{-\eta}\right. \\
&\left. + a(x)\left(A_0' + B_0'e^{-\eta}\right) + g(x, A_0 + B_0 e^{-\eta})\right) = 0.
\end{aligned} \tag{5.45}$$

We expand

$$g(x, A_0 + B_0 e^{-\eta}) = g(x, A_0) + g_y(x, A_0)B_0 e^{-\eta} + \frac{1}{2}g_{yy}(x, A_0)(B_0^2 e^{-2\eta}) + \ldots$$

about $y = A_0$. To prevent secular terms in Y_1, we must make the coefficients of 1 and $e^{-\eta}$ in the forcing term of (5.45) be zero. Thus, we will need A_0 to satisfy the limiting nonlinear equation

$$a(x)A_0' + g(x, A_0) = 0$$

while B_0 must satisfy the coupled linear equation

$$-a(x)B_0' + (-a'(x) + g_y(x, A_0))B_0 = 0.$$

Using the terminal condition for A_0, we shall assume that a unique solution to the reduced problem

$$A_0' = -\frac{g(x, A_0)}{a(x)}, \quad A_0(1) = y(1) \tag{5.46}$$

is defined throughout $0 \leq x \leq 1$. We may have to obtain A_0 numerically. Then, $B_0(x)$ is uniquely determined from the linear problem

$$(aB_0)' = g_y(x, A_0)B_0, \quad B_0(0) = y(0) - A_0(0),$$

i.e.

$$B_0(x) = e^{\int_0^x \frac{g_y(s, A_0(s))}{a(o)} ds} \frac{a(0)}{a(x)} (y(0) - A_0(0)). \tag{5.47}$$

This completely specifies Y_0. Next, we will need to integrate

$$Y_{1\eta\eta} + Y_{1\eta} + \frac{1}{a^2(x)}(g(x, A_0 + B_0 e^{-\eta}) - g(x, A_0) - g_y(x, A_0)B_0 e^{-\eta}) = 0$$

using variation of parameters to determine its complementary solution by applying the initial conditions and then eliminating the resulting resonant terms in the differential equation for Y_2. Generalizations of these methods are found in O'Malley [362, 363], Smith [465], and elsewhere.

To illustrate how our earlier results on turning point problems fit the two-time ansatz, consider the two examples that follow.

Example 1

Recall that solutions of the two-point problem

$$\epsilon y'' + xy' = 0, \quad -1 \le x \le 1, \quad y(\pm 1) = \pm 1 \tag{5.48}$$

have the form

$$y(x, \epsilon) = A + B \int_0^x e^{-s^2/2\epsilon} ds$$

for constants $A(\epsilon)$ and $B(\epsilon)$. Applying the boundary conditions, we get $A = 0$ and $B = (\int_0^1 e^{-s^2/2\epsilon} ds)^{-1}$, so the asymptotic solution

$$y(x, \epsilon) \sim \frac{\int_0^{x/\sqrt{\epsilon}} e^{-t^2/2} dt}{\int_0^\infty e^{-t^2/2} dt} \tag{5.49}$$

is an odd function of the stretched variable

$$\xi = x/\sqrt{\epsilon}. \tag{5.50}$$

If we directly sought the solution as

$$y(x, \epsilon) = C(\xi), \tag{5.51}$$

C would satisfy the boundary value problem

$$\frac{d^2 C}{d\xi^2} + \xi \frac{dC}{d\xi} = 0, \quad C(\pm\infty) = \pm 1,$$

as found. The symmetric shock layer $C(\xi)$ clearly connects the outer solutions ∓ 1 on opposite sides of the turning point.

Example 2

Recall that the differential equation

$$\epsilon y'' + 2xy' - 2y = 0, \quad -1 \leq x \leq 1, \quad y(-1) = -1, \quad y(1) = 2 \qquad (5.52)$$

has the exact solution

$$y(x, \epsilon) = x + x \frac{\int_{-1}^{x} e^{-s^2/\epsilon}\, ds}{\int_{-1}^{1} e^{-s^2/\epsilon}\, ds} + \frac{\epsilon e^{-x^2/\epsilon}}{2 \int_{-1}^{1} e^{-s^2/\epsilon}\, ds}$$

depending on the variables x and $\xi \equiv x/\sqrt{\epsilon}$. Indeed, it has the simple form

$$y(x, \epsilon) = Y(x, \xi, \sqrt{\epsilon}) \sim x + xC_0(\xi) + \epsilon C_2(\xi) \qquad (5.53)$$

where $C_0(\xi) \equiv \frac{\int_{-\infty}^{\xi} e^{-t^2}\, dt}{\int_{-\infty}^{\infty} e^{-t^2}\, dt}$ and $C_2(\xi) \equiv \frac{e^{-\xi^2}}{2 \int_{-\infty}^{\infty} e^{-t^2}\, dt}$. As expected,

$$Y(x, \xi, 0) \to \begin{cases} 2x & \text{as } \xi \to \infty \\ x & \text{as } \xi \to -\infty \end{cases}$$

while $C_2(\xi) \to 0$ as $\xi \to \pm\infty$.

Murdock [335] considered the so-called *harmonic resonance* problem of finding solutions of period $\frac{2\pi}{\omega(\epsilon)}$ for the equation

$$\ddot{y} + y = \epsilon f(y, \dot{y}, \omega(\epsilon)t) \qquad (5.54)$$

where

$$\omega(\epsilon) \sim 1 + \epsilon\omega_1 + \epsilon^2\omega_2 + \dots \qquad (5.55)$$

is specified. He lets the initial values take the form

$$\begin{cases} y(0) & = \alpha_0 + \epsilon\alpha_1 + \dots \\ \text{and} \\ \dot{y}(0) & = \beta_0 + \epsilon\beta_1 + \dots \end{cases} \qquad (5.56)$$

(to be determined termwise) with

$$y \sim y_0(\omega t) + \epsilon y_1(\omega t) + \dots. \qquad (5.57)$$

Exercises

1. To further motivate two-scale expansions, consider the scalar linear initial value problem

$$\epsilon u' + a(x)u = b(x), \quad x \geq 0, \quad u(0) \quad \text{prescribed}$$

on a finite interval where $a(x) > 0$ and a and b are smooth.

(a) Obtain a formal asymptotic solution in the form

$$u(x, \epsilon) = U(x, \epsilon) + e^{-\frac{1}{\epsilon} \int_0^x a(s)\, ds} (u(0) - U(0, \epsilon))$$

where U is an outer expansion

$$U(x, \epsilon) \sim U_0(x) + \epsilon U_1(x) + \epsilon^2 U_2(x) + \dots$$

(b) Integrate the exact solution

$$u(x, \epsilon) = e^{-\frac{1}{\epsilon} \int_0^x a(s)\, ds} u(0) + \frac{1}{\epsilon} \int_0^x e^{-\frac{1}{\epsilon} \int_t^x a(s)\, ds} b(t)\, dt$$

by parts to show that

$$u(x, \epsilon) = \frac{b(x)}{a(x)} + e^{-\frac{1}{\epsilon} \int_0^x a(s)\, ds} \left(u(0) - \frac{b(0)}{a(0)} \right) + O(\epsilon).$$

(c) Integrate the exact solution again by parts to show that

$$u(x, \epsilon) = U_0(x) + \epsilon U_1(x) + e^{-\frac{1}{\epsilon} \int_0^x a(s)\, ds} (u(0) - U_0(0) - \epsilon U_1(0)) + O(\epsilon^2).$$

(d) Find the exact solution to the two-point boundary value problem

$$\epsilon y'' + a(x)y' = f(x), \quad 0 \leq x \leq 1$$

with

$$y(0) \quad \text{and} \quad y(1) \text{ prescribed.}$$

For $a(x) > 0$ and a and f smooth, use integration by parts to show that the asymptotic solution has the two-variable form

$$y(x, \epsilon) = Y(x, \epsilon) + e^{-\frac{1}{\epsilon} \int_0^x a(s)\, ds} (y(0) - Y(0, \epsilon))$$

where the outer solution Y has an asymptotic series expansion in ϵ.

2. Consider the scalar two-point problem

$$\epsilon y'' + f(x, y, y', \epsilon) = 0, \quad 0 \leq x \leq 1$$

with $y(0)$ and $y(1)$ prescribed in cases when the reduced problem

$$f(x, Y, Y', 0) = 0, \quad Y(1) = y(1)$$

has a solution $Y_0(x)$ on $0 \leq x \leq 1$ with

$$f_{y'}(x, Y_0, Y_0', 0) \geq \sigma$$

for a positive constant σ. Provide examples for which one can use the fast variable

$$\frac{1}{\epsilon} \int_0^t f_{y'}(s, Y_0(s), Y_0'(s), 0)\, ds$$

(cf. Willett [524], O'Malley [362], Searl [443], and Rosenblat [420]).

3. (cf. Searl [443]) Consider the *Cole-Lagerstrom problem*

$$\epsilon\ddot{x} + x\dot{x} - x = 0, \qquad x(0) = \alpha, \quad x(1) = \beta.$$

(a) Try solving the problem via two-timing by setting

$$x(t, \tau, \epsilon) = x_0(t, \tau) + \epsilon x_1(t, \tau) + \cdots$$

with the slow time $\tau = \epsilon t$. Show that x_0 must satisfy

$$x_{0\tau\tau} + x_0 x_{0\tau} = 0$$

while

$$x_{1\tau\tau} + x_0 x_{1\tau} + x_1 x_{0\tau} + 2x_{0t\tau} + x_0 x_{0\tau} - x_0 = 0.$$

Then take

$$x_0(t, \tau) = u_0(t) \tanh\left(\frac{u_0(t)}{2}\tau + v_0(t)\right)$$

with $u_0(1) = \beta$ and $v_0(0) = \tanh^{-1}\left(\frac{\alpha}{u_0(0)}\right)$.

(b) Determine the functions u_0 and v_0 to eliminate secular terms in x_1. Note that $\operatorname{sech}^2\left(\frac{u_0}{2}\tau + v_0\right)$ is a solution of the homogeneous linearized equation

$$x_{\tau\tau} + x_0(t, \tau)x_\tau + x_{0\tau}(t, \tau)x = 0,$$

with t as a parameter. Searl determines $u_0(t) = t + \beta - 1$ and $v_0(t) = v_0(0) = \tanh^{-1}\left(\frac{\alpha}{\beta-1}\right)$ when $|\alpha| < |\beta - 1|$.

(c) Can you find solutions for all αs and βs?

4. (a) Solve the linear initial value problem

$$\ddot{y} + y + \epsilon e^{-t}y = 0, \quad y(0) = 1, \quad \dot{y}(0) = 0$$

on $t \leq 0$ in terms of Bessel functions.

(b) Show that the regular perturbation expansion provides the asymptotic solution with no secular terms.

The Palestinian-American Ali Nayfeh (1933–) got his Ph.D. in aeronautics at Stanford in 1964, with Milton Van Dyke as his advisor. He's been in the department of engineering science and mechanics at Virginia Tech since 1971. His 1973 text, *Perturbation Methods* [341], now reissued as a Wiley Classic, surveyed the growing literature and provided detailed solutions to numerous examples. This and several related books by him have been very successful pedagogically for two generations of engineering and science students. Nayfeh [342] again provides solutions to many perturbation problems, while Nayfeh [343] updates his discussion of two-timing.

(b) Lighthill's Method

Sir M. James Lighthill (1924–1988) was a British applied mathematician and administrator who held important positions at the University of Manchester, the Royal Aircraft Establishment, Imperial College London, Trinity College Cambridge, and University College London (see Pedley [300] and the biography Debnath [115]).

One of the topics presented by Nayfeh [341] is *coordinate stretching*. It generalizes the Poincaré-Lindstedt method and was called the *PLK method* by von Kármán's student H.-S. Tsien (after Poincaré, Lighthill, and Kuo) (cf. Tsien [486]). (Tsien lost his security clearance in 1950 and spent 5 years under house arrest in California before returning to China to lead its rocket program. He is the subject of a biography (Chang [75]). Lighthill gave the *Ludwig Prandtl Memorial Lecture* to GAMM in 1961. He commented

> Indeed, his revolutionary discovery of the boundary layer in 1904
> had the same transforming effect on fluid mechanics as Einstein's
> 1905 discoveries on other parts of physics.

A simple example of Lighthill's method (cf. Lighthill [290], Nayfeh [341], de Jager and Jiang [224], and Johnson [226]) is provided by the nonlinear initial value problem

$$(x + \epsilon u)\frac{du}{dx} + u = 0, \quad u(1) = 1. \tag{5.58}$$

A regular perturbation expansion

$$u(x, \epsilon) = u_0(x) + \epsilon u_1(x) + \dots$$

breaks down at $x = 0$, though the given equation only becomes singular when $x + \epsilon u = 0$. Proceeding termwise, we'd need

$$x\frac{du_0}{dx} + u_0 = 0, \quad u_0(1) = 1,$$

so $u_0(x) = 1/x$. Then $x\frac{du_1}{dx} + u_1 + u_0\frac{du_0}{dx} = 0$ and $u_1(1) = 0$ imply that

$$u_1(x) = \frac{x^2 - 1}{2x^3}.$$

The increasing singularity of the terms $u_k(x)$ at $x = 0$ is a difficulty which we might compensate for by introducing a *near-identity transformation*

$$x = \xi + \epsilon f(\xi, \epsilon) \sim \xi + \epsilon f_0(\xi) + \epsilon^2 f_1(\xi) + \dots \tag{5.59}$$

with a yet unspecified function f to define the new coordinate $\xi(x, \epsilon)$ by inversion and a corresponding regular perturbation expansion

$$U(\xi, \epsilon) = U_0(\xi) + \epsilon U_1(\xi) + \dots \tag{5.60}$$

for the solution of (5.58) as a function of ξ, which we hope will be defined at $x = 0$. (Bush [62] makes the helpful suggestion that subsequent coefficients U_k should be no more singular than previous ones.) Since $\frac{d\xi}{dx} = \frac{1}{1+\epsilon \frac{df}{d\xi}}$, the given differential equation (5.58) transforms to

$$(\xi + \epsilon f(\xi, \epsilon) + \epsilon U)\frac{dU}{d\xi} + \left(1 + \epsilon \frac{df}{d\xi}(\xi, \epsilon)\right) U = 0. \tag{5.61}$$

The regular perturbation process now implies the sequence of equations

$$\xi \frac{dU_0}{d\xi} + U_0 = 0, \tag{5.62}$$

$$\xi \frac{dU_1}{d\xi} + U_1 + f_0(\xi)\frac{dU_0}{d\xi} + \frac{df_0}{d\xi}U_0 + U_0\frac{dU_0}{d\xi} = 0, \tag{5.63}$$

etc., for the coefficients U_k in (5.60). The boundary values need to be determined from the terminal condition

$$U(\xi^*, \epsilon) = 1 \quad \text{where} \quad 1 = \xi^* + \epsilon f(\xi^*, \epsilon). \tag{5.64}$$

Taking

$$\xi^* \sim 1 + b_0\epsilon + b_1\epsilon^2 + \ldots, \tag{5.65}$$

the ϵ coefficient in (5.64), $1 = (1 + b_0\epsilon + b_1\epsilon^2 + \ldots) + \epsilon f_0(1 + \epsilon b_0 + \ldots) + \epsilon^2 f_1(1 + \ldots) + \ldots$, implies that

$$b_0 = -f_0(1),$$

so $U(\xi^*, \epsilon) = U_0(1 - f_0(1)\epsilon + \ldots) + \epsilon U_1(1 + \ldots) + \ldots = 1$ determines the end values

$$U_0(1) = 1, \tag{5.66}$$

and

$$U_1(1) = U_0'(1)f_0(1) \tag{5.67}$$

etc. Returning to (5.62), $\frac{d}{d\xi}(\xi U_0) = 0$, $U_0(1) = 1$ implies that

$$U_0(\xi) = \frac{1}{\xi} \tag{5.68}$$

(compared to $1/x$ for u_0). Now there still remains much flexibility in picking f in the near-identity transformation (5.59). We will compensate the singular term $U_0\frac{dU_0}{d\xi}$ in (5.63) by asking that f_0 satisfies

$$\frac{df_0}{d\xi}U_0 + f_0\frac{dU_0}{d\xi} + U_0\frac{dU_0}{d\xi} = 0, \tag{5.69}$$

leaving

$$\xi\frac{dU_1}{d\xi} + U_1 = 0, \quad U_1(1) = -f_0(1). \tag{5.70}$$

from (5.63). If we now take $f_0(1) = 0$, we get

$$U_1(\xi) = 0, \tag{5.71}$$

leaving (5.63) as the initial value problem $\frac{1}{\xi}\frac{df_0}{d\xi} - \frac{f_0}{\xi^2} - \frac{1}{\xi^3} = 0$, $f_0(1) = 0$. Integration yields

$$f_0(\xi) = \frac{1}{2}\left(\xi - \frac{1}{\xi}\right). \tag{5.72}$$

Taking all later f_ks in (5.59) to be zero, as well as all later U_ks in (5.60), we simply obtain the quadratic near-identity transformation

$$x = \xi + \frac{\epsilon}{2}\left(\xi - \frac{1}{\xi}\right) \tag{5.73}$$

and the one-term solution

$$U = U_0(\xi) = \frac{1}{\xi}. \tag{5.74}$$

Since (5.73) has the inverse $\xi = \frac{x+\sqrt{x^2+2\epsilon+\epsilon^2}}{2+\epsilon}$, the solution (5.74) of (5.58) is

$$u(x, \epsilon) = \frac{2+\epsilon}{x + \sqrt{x^2 + 2\epsilon + \epsilon^2}} = \frac{-x + \sqrt{x^2 + 2\epsilon + \epsilon^2}}{\epsilon}. \tag{5.75}$$

Amazingly, this is the *exact* solution, as can be checked by integrating

$$\left(xu + \frac{\epsilon}{2}u^2\right)' = 0, \quad u(1) = 1.$$

We note a closely related method of George Temple [482]. Recall, too, that Kaplun [235] called a coordinate ξ *optimal* when it leads to a uniformly valid solution $U(\xi, \epsilon)$. Readers should consult Comstock [95] regarding Lighthill's method and the controversy that once surrounded it. Johnson [226] considers more general initial value problems for

$$(x + \epsilon u)u' + (\alpha + \beta x)u = 0$$

for $\alpha > 0$ and $u(1)$ prescribed, while Sibuya and Takahasi [459] provide a proof for equations

$$(x + \epsilon u)u' + q(x)u = r(x).$$

Awrejcewicz and Krysko [18] more generally suppose one begins with a *naive* expansion

$$f(x, \epsilon) \sim f_0(x) + \epsilon f_1(x) + \epsilon^2 f_2(x) + \dots \tag{5.76}$$

that is not uniformly valid. They next introduce the deformed variable X via

$$x = X + \epsilon \nu_1(X) + \epsilon^2 \nu_2(X) + \dots \qquad (5.77)$$

to obtain

$$f(x, \epsilon) = F(X, \epsilon) \sim F_0(X) + \epsilon F_1(X) + \epsilon^2 F_2(X) + \dots. \qquad (5.78)$$

Then, they pick the deformation coefficients ν_1, ν_2, ... in (5.77) to achieve a uniformly suitable series for F.

(c)   Phase-Plane Methods and Relaxation Oscillations

Scalar boundary value problems for singularly perturbed equations in *conservation form*,

$$\begin{cases} \epsilon^2 \ddot{x} + f(x) = 0, \quad 0 \le t \le 1 \\ \text{with} \\ x(0) \text{ and } x(1) \text{ prescribed,} \end{cases} \qquad (5.79)$$

can be integrated by introducing the potential energy

$$V(x) = \int^x f(s) \, ds \qquad (5.80)$$

and invoking the resulting *conservation of energy* principle

$$\frac{1}{2}\epsilon^2 \dot{x}^2 + V(x) = E \qquad (5.81)$$

for a constant total energy E fixed on each solution trajectory (cf. O'Malley [367], Lutz and Goze [300], and Ou and Wong [384]). (The classical graphical treatment of conservation equations with $\epsilon = 1$ is given, e.g., in Jordan and Smith [230].) This follows immediately by direct integration, after multiplying the differential equation of (5.79) by $\dot{x}$. If we set

$$y = \epsilon \dot{x}, \qquad (5.82)$$

we can describe the motion in the x-y *phase plane* by considering the singularly perturbed system

$$\begin{cases} \epsilon \dot{x} = y \\ \epsilon \dot{y} = -f(x). \end{cases} \qquad (5.83)$$

To get a real trajectory, the constant E must always exceed $V(x)$ since $E - V = y^2/2 \ge 0$.

Indeed, since $dt = \pm \frac{\epsilon \, dx}{\sqrt{2(E - V(x))}}$, E must be only just slightly greater than the maximum of V on any bounded trajectory in order to use up one unit of transit time in going from one prescribed endvalue to the other, i.e., so that

$$\int_0^1 dt = \epsilon \int_{x(0)}^{x(1)} \frac{dx}{\sqrt{2(E - V(x))}} = 1 \tag{5.84}$$

for integration along the path $x(t)$ traversed. Moreover, most time must be spent near rest points of (5.83) corresponding to such maxima, because $dt = O(\epsilon)$ elsewhere. Be aware that such two-point problems generally have more than one solution that follow related, but different, phase-plane trajectories (with slightly different E levels), as we shall demonstrate with Example 2 below.

Example 1

Consider the simple linear problem

$$\begin{cases} \epsilon^2 \ddot{x} - x = 0, & 0 \le t \le 1 \\ \text{with} \\ x(0) = 1 \text{ and } x(1) = 2. \end{cases} \tag{5.85}$$

Here, we can take

$$V = \frac{1}{2}x^2.$$

Note that solutions $x(t)$ satisfy a maximum principle (cf. Dorr et al. [124]). Since solutions are linear combinations of $e^{\pm t/\epsilon}$, we write the unique solution of (5.85) as

$$x(t, \epsilon) = e^{-t/\epsilon}c + e^{-(1-t)/\epsilon}k$$

for constants $c(\epsilon)$ and $k(\epsilon)$ that satisfy the linear system

$$c + e^{-1/\epsilon}k = 1 \quad \text{and} \quad e^{-1/\epsilon}c + k = 2.$$

Up to asymptotically negligible quantities, we get $c \sim 1$ and $k \sim 2$, so

$$x(t, \epsilon) \sim e^{-t/\epsilon} + 2e^{-(1-t)/\epsilon}. \tag{5.86}$$

Then

$$y = \epsilon\dot{x} \sim -e^{-t/\epsilon} + 2e^{-(1-t)/\epsilon} \tag{5.87}$$

and the total energy on the corresponding trajectory is

$$E = \frac{1}{2}(y^2 - x^2) \sim -4e^{-1/\epsilon} \tag{5.88}$$

(negative, but asymptotically negligible). Graphically, see Figs. 5.1 and 5.2.

We plot $V(x)$ with a small negative E value in Fig. 5.3. This determines the allowed range of x values (omitting a neighborhood of $x = 0$) and graphically determines the corresponding real-valued $y = \pm\sqrt{2(E - V(x))}$.

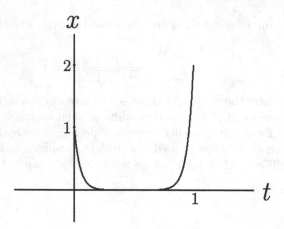

Figure 5.1: The asymptotic solution $x(t, \epsilon)$ of $\epsilon^2 \ddot{x} = x$

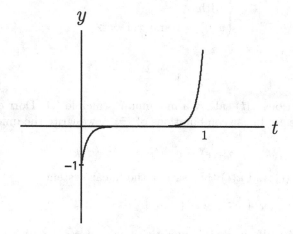

Figure 5.2: The solution $y(t, \epsilon) = \epsilon \dot{x}(t, \epsilon)$ of $\epsilon^2 \ddot{x} = x$

To obtain a trajectory joining $x(0) = 1$ and $x(1) = 2$, we must follow the dashed right orbit in the phase-plane (Fig. 5.4) where $x > 0$ and y is monotonically increasing. Let α' and α be the points where the orbit cuts the vertical line $x = 1$ (with α' below α) and let β' and β be the corresponding points where it cuts $x = 2$. Because motion only slows down near the rest point $(0, 0)$, $\epsilon \int_1^2 \frac{dx}{y} = 1$ requires y to be small most of the time. This rules out the trajectory $\alpha\beta$ as being too fast, so the only possible orbit is $\alpha'\alpha\beta$, which moves slowly near the rest point, but fast in the endpoint layers, as pictured in Figs. 5.1 and 5.2.

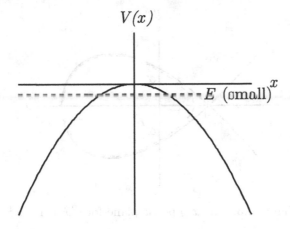

Figure 5.3: The potential $V(x) = -\frac{1}{2}x^2$

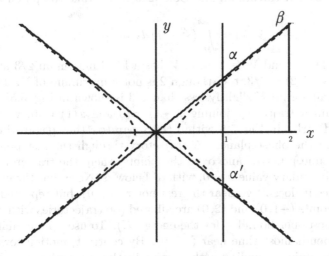

Figure 5.4: The dotted phase-plane orbit for the solution $\alpha'\alpha\beta$ of $\epsilon^2\ddot{x} = x$ with ϵ small

Example 2

Consider the nonlinear example

$$\begin{cases} \epsilon^2\ddot{x} + x^2 = 1, & -1 \le t \le 1 \\ \text{with} \quad x(\pm 1) = 0. \end{cases} \tag{5.89}$$

The example is important because it suggests that the method of matched expansions can mislead by suggesting the existence of *spurious solutions*. Actual solutions can be obtained in terms of elliptic integrals (cf. Byrd and

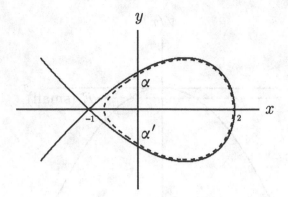

Figure 5.5: The x-y phase plane for $\epsilon^2 \ddot{x} = 1 - x^2$

Friedman [66] and Kevorkian and Cole [249]). We take the potential energy to be

$$V(x) = \int_0^x (s^2 - 1)\, ds = \frac{x^3}{3} - x. \tag{5.90}$$

Since $V'(-1) = 0$ and $V''(-1) < 0$, V has a local maximum $2/3$ at $x = -1$. Significantly, $V(2) = 2/3$, too, though 2 is not a maximum of V. The phase-plane portrait for an E slightly less than $2/3$ is shown in Fig. 5.5.

To obtain a trajectory joining $x(-1) = 0$ and $x(1) = 0$, we will need the (dashed) orbit in Fig. 5.5 within the *separatrix* that passes through the rest point in the phase-plane. (A trajectory through the rest point cannot take finite time.) Let α' and α be the points where the trajectory hits the prescribed boundary value $x = 0$, with α' below α. Note that the orbit $\alpha\alpha'$ is too fast since it doesn't go near the rest point $(-1, 0)$, but repeated passages past both points $(-1, 0)$ and $(2, 0)$ are allowed (on trajectories with somewhat less than the upper bound $\frac{2}{3}$ for the energy E). To use up one unit of time, the orbit spends most time near $(-1, 0)$. By contrast, motion toward $(2, 0)$ and back is rapid, providing a thin *spike* in the x-y trajectory. The limit $X_0(t) = -1$ is a root of the reduced equation. The other root, $X_0(t) = 1$, a minimum of the potential energy V, is quite irrelevant asymptotically. The simplest (and shortest) solution $\alpha'\alpha$ is shown in Fig. 5.6. It has the form

$$x(t, \epsilon) \sim -1 + \frac{12 e^{p_L}}{(1 + e^{p_L})^2} + \frac{12 e^{p_R}}{(1 + e^{p_R})^2} \tag{5.91}$$

(cf. Carrier and Pearson [71] and Lange [280]) where $p_{L/R} = \frac{\sqrt{2}}{\epsilon}(1 \pm t) + 2\ln(\sqrt{3} + \sqrt{2})$ or, equivalently,

$$
\begin{aligned}
x(t, \epsilon) \sim\ & -1 + 3\operatorname{sech}^2\left(\frac{1+t}{\sqrt{2}\epsilon} + \ln(\sqrt{3} + \sqrt{2})\right) \\
& + 3\operatorname{sech}^2\left(\frac{1-t}{\sqrt{2}\epsilon} + \ln(\sqrt{3} + \sqrt{2})\right).
\end{aligned}
\tag{5.92}
$$

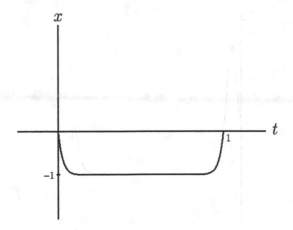

Figure 5.6: The x trajectory following $\alpha'\alpha$

The corresponding solution for $y = \epsilon \dot{x}$ follows by differentiation. Readers are urged to plot these functions to check the complicated formulas. Other endpoint layers, with spikes to 2, are possible for trajectories with lower energies E. For example, the solution $\alpha\alpha'\alpha$ has such an initial spike while $\alpha'\alpha\alpha'$ has a terminal spike, and $\alpha\alpha'\alpha\alpha'$ has a spike near both endpoints. Two of these solutions are shown in Figs. 5.7 and 5.8. Note the differences in the signs and sizes of the endpoint slopes.

According to Ou and Wong [384], the asymptotic solutions are, respectively, given by

$$x(t,\epsilon) \sim -1 + 3\,\mathrm{sech}^2\left(-\frac{t+1}{\sqrt{2}\epsilon} + \ln(\sqrt{3}+\sqrt{2})\right)$$
$$+ 3\,\mathrm{sech}^2\left(\frac{1-t}{\sqrt{2}\epsilon} + \ln(\sqrt{3}+\sqrt{2})\right) \tag{5.93}$$

$$x(t,\epsilon) \sim -1 + 3\,\mathrm{sech}^2\left(\frac{t+1}{\sqrt{2}\epsilon} + \ln(\sqrt{3}+\sqrt{2})\right)$$
$$+ 3\,\mathrm{sech}^2\left(-\frac{(1-t)}{\sqrt{2}\epsilon} + \ln(\sqrt{3}+\sqrt{2})\right), \tag{5.94}$$

and

$$x(t,\epsilon) \sim -1 + 3\,\mathrm{sech}^2\left(-\frac{(t+1)}{\sqrt{2}\epsilon} + \ln(\sqrt{3}+\sqrt{2})\right)$$
$$+ 3\,\mathrm{sech}^2\left(-\frac{(1-t)}{\sqrt{2}\epsilon} + \ln(\sqrt{3}+\sqrt{2})\right). \tag{5.95}$$

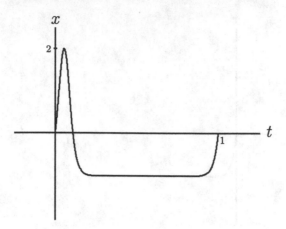

Figure 5.7: The x trajectory following $\alpha\alpha'\alpha$

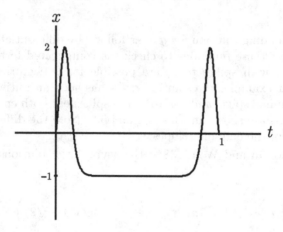

Figure 5.8: The x trajectory following $\alpha\alpha'\alpha\alpha'$

Interior spikes are another possibility. Carrier and Pearson [71] warned that classical matching allows one to formally add a spike

$$g(t) = 3\text{sech}^2\left(\frac{t - t_0}{\sqrt{2}\epsilon}\right) \qquad (5.96)$$

about *any* interior point t_0 to obtain a possibly new asymptotic solution x. Indeed, one could seem to add several isolated spikes. However, Carrier and Pearson realized by a phase-plane analysis (such as ours) that these "solutions" are generally *spurious*. It would be allowable to add a single spike at the midpoint $t_0 = 0$ or two such spikes simultaneously at $t_0 = \pm 1/3$ (corresponding to one- and two-thirds of the interval length). Solutions with legitimate interior spikes can be obtained that pass near the rest point more than once, on a longer trajectory with somewhat less energy E. Such cycles take the same time of passage for each revolution in the phase- plane, so the resulting periodic motion in the x-t plane must feature nearly regularly spaced spikes, in addition to the endpoint layers already considered. Carrier and Pearson reassured readers

> The authors have never seen this occur with any problem which arose in a scientific context.

Nonetheless, our confidence in formal matching is diminished. Kath [242] provides extensions to slowly varying phase-planes.

Limiting attention to N *regularly spaced* $O(\epsilon)$-thick interior spikes will, as N increases, ultimately fill the t interval $(-1, 1)$, leaving scant space for the attractive outer limit -1 to apply. Thus, it's not surprising that Ou and Wong [384], using a natural *shooting* argument for initial values $x(-1, 0) = 0$ and a varying $\dot{x}(-1, 0) = k$, were able to show that actual solutions for appropriate k have at most $O(1/\epsilon)$ internal spikes. They also provided details regarding the asymptotic locations of those spikes as functions of ϵ. Also, note Ward [510].

Carrier [72] considered the nonautonomous problem

$$\epsilon^2 \ddot{x} + 2(1 - t^2)x + x^2 - 1 = 0, \quad x(\pm 1) = 0. \tag{5.97}$$

No phase-plane argument applies. As you'd expect, however, he finds (cf. Bender and Orszag [36]) that the asymptotic solutions have the outer limit

$$-(1 - t^2) - \sqrt{(1 - t^2)^2 + 1}$$

with both boundary layers and various additional spikes possible. MacGill-ivray et al. [303] show that the interior spikes coalesce as $\epsilon \to 0$. Carrier again raised the specter of spurious solutions and considered the possibility of multiple solutions, differing in their endpoint slopes by only negligible amounts. Bender and Orszag display a number of numerical solutions for $\epsilon = 10^{-4}$, again suggesting a breakdown in the asymptotics as the number of spikes increases. The more recent results of Ai [4] suggest a pileup of spikes near the midpoint, but that isn't clear from Wong and Zhao [527], who also used shooting arguments. Hastings and McLeod [198] transform the problem to a more tractable equation of the Riccati form

$$\epsilon^2 \ddot{u} = u(q(t, \epsilon) - u)$$

by setting $x = X + u$ for the outer solution X (i.e., by using the subtraction trick). They also develop extensive material regarding spikes and layers for more general reaction-diffusion models.

Carrier and Pearson [71] use singular perturbations as the penultimate topic in their ODE text. George Carrier (1918–2002) was a superb math modeller who repeatedly used asymptotic techniques to comprehend a wide variety of physical applications. He was active as a consultant to industry and government, winning the National Medal of Science in 1990. He also had great success and influence as a teacher, throughout his long career at Harvard (from 1952). Even earlier, he introduced Julian Cole to perturbation methods as an undergraduate at Cornell (where Carrier got his Ph.D.). Carl Pearson co-authored several books with Carrier, worked for Boeing, and was a professor of aeronautics and applied mathematics at the University of Washington.

Let's next consider an autonomous slow-fast planar system

$$\begin{cases} \dot{x} = f(x,y) \\ \epsilon\dot{y} = g(x,y) \end{cases} \tag{5.98}$$

for times $t \geq 0$. We can expect slow motion to follow the reduced system

$$\begin{cases} \frac{dX}{dt} = f(X,Y) \\ 0 = g(X,Y), \end{cases} \tag{5.99}$$

while fast motion should follow the stretched system

$$\frac{dy}{d\tau} = g(x,y) \tag{5.100}$$

with x as a parameter and fast time

$$\tau = \frac{t - t_0}{\epsilon} \quad \text{for some } t_0. \tag{5.101}$$

Thus, slow motion will lie on the *manifold*

$$\Gamma : g(x,y) = 0 \tag{5.102}$$

and fast motion will be off it. If the prescribed initial point

$$(x(0), y(0))$$

is not on Γ, we can immediately expect nearly vertical fast motion toward (or away from) Γ since g/ϵ will be large. If we next reach a stable (i.e., attractive) point

$$(x(0), y_1)$$

on Γ where $g_y < 0$, we then expect slow motion along Γ to follow until, say, g_y loses stability at a *junction point*

$$(x_2, y_2).$$

Then, we can again expect rapid motion away from the manifold until, perhaps, a stable *drop point*

$$(x_2, y_3)$$

on Γ is reached, when slow motion may again begin. When successive alternations between slow and fast motions produce a limiting *closed trajectory* with jerky, almost instantaneous, jumps in y, we will say we have a *relaxation oscillation*. The limiting period will be determined by integrating

$$dt = \frac{dx}{f(x,y)}$$

on the slow manifold Γ. Detailed asymptotics, especially near junction and drop points, is called for. We leave the connection to *hysteresis* open, but interested readers might note Mortell et al. [328].

Many times, oscillators like (5.98) arise from scalar second-order differential equations

$$\epsilon\ddot{y} - F'(y)\dot{y} + y = 0. \tag{5.103}$$

The van der Pol equation occurs when

$$F(y) = y - \frac{1}{3}y^3, \tag{5.104}$$

i.e. from

$$\frac{d^2y}{d\tau^2} - \lambda(1 - y^2)\frac{dy}{d\tau} + y = 0 \tag{5.105}$$

when $\lambda = \frac{1}{\sqrt{\epsilon}}$ is large and $\tau = \lambda t$. For it, we let

$$\dot{x} = f(x,y) \equiv y \tag{5.106}$$

and integrate the resulting y equation (5.103) to get

$$\epsilon\dot{y} = g(x,y) \equiv F(y) - x. \tag{5.107}$$

Thus,

$$\Gamma : x = F(y) \tag{5.108}$$

is then an S-shaped curve, stable for $|y| > 1$, so the relaxation oscillation for the van der Pol equation jumps between arcs of Γ at $y = \pm 1$ (cf. Stoker [476]). See Figs. 5.9 and 5.10.

As anticipated, the limiting period of the corresponding trajectory will be

$$T = 2\int_{-2/3}^{2/3} \frac{dx}{y(x)} = 2\int_{2}^{1} \frac{F'(y)}{y}dy = 3 - 2\ln 2, \tag{5.109}$$

integrating along one arc of Γ. Obtaining higher-order terms in the asymptotic expansion for the period or the amplitude requires much effort (cf. Stoker [476], Levinson [287], Mischenko and Rozov [321], and Grasman [179]). In particular, the asymptotic sequences that arise are far from obvious *a priori*.

The first to study relaxation oscillations seems to be Balthasar van der Pol (1889–1959) in 1926 (cf. Israel [222]). He was a scholarly Dutch engineer, with a Ph.D. from Utretcht based on work done in Cambridge, and was head physicist at Philips Physical Laboratory. He became particularly interested in modeling the human heart and its arrhythmias. (From 1945 to 1946, he was president of the Temporary University at Eindhoven, founded to replace other Dutch universities in occupied territory.) Currently, applications to neuroscience, featuring many coupled oscillations, are of great interest and quite complicated (cf. Ermentrout and Terman [142]). Earlier neural networks are modeled in Cronin [105, 106]. The simplest examples may be see-saws with water reservoirs, pictured in Grasman [179]. For other applications, see Sastry and Desoer [432].

One of the most interesting recent developments relates to the occurrence of *canards*. The topic was introduced by a group of French mathematicians in the 1970s, primarily concerned with applying non-standard analysis (cf. Diener and Diener [120]). (Some were working in Algeria.) They considered the forced van der Pol equation

$$\epsilon \ddot{y} + (y^2 - 1)\dot{y} + y = a \tag{5.110}$$

or, equivalently, the system

$$\begin{cases} \epsilon \dot{y} &= z - F(y) \\ \dot{z} &= a - y \end{cases} \tag{5.111}$$

in the Liénard plane where $F(y) = y - \frac{y^3}{3}$ and a is a constant. For $a = 0$ or 1, for example, we get periodic solutions that follow the limit cycle consisting of slow motion on the attractive arcs of the characteristic curve and fast

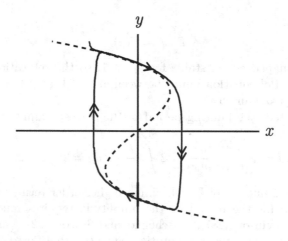

Figure 5.9: Approaching the limit cycle for the van der Pol equation

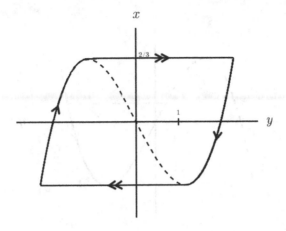

Figure 5.10: Limit cycle

horizontal trajectories. They found that canards occur for special values

$$a(\epsilon) \sim a_0 + \epsilon a_1 + \epsilon^2 a_2 + \ldots \qquad (5.112)$$

near $a_0 = 1$, resulting in solutions of the form

$$z \sim F(y) + \epsilon f_1(y) + \epsilon^2 f_2(y) + \ldots \qquad (5.113)$$

that travel up the unstable branch of the characteristic curve to produce a *canard sans tête* (Fig. 5.11) or a *canard avec tête* (Fig. 5.12). (Pardon my French!) Note the connection to Sect. (c) of Chap. 4. Motion up the dashed curve is unstable (i.e., repulsive).

Finding the power series is straightforward since $\dot{z} = z'\dot{y}$ and $\epsilon\dot{y} = z - F(y)$ imply that

$$\dot{z} = a_0 - y + \epsilon a_1 + \ldots = (F'(y) + \epsilon f_1'(y) + \ldots)(f_1(y) + \epsilon f_2(y) + \ldots).$$

Thus, $F'(1) = 0$ shows that we need

$$a_0 = 1. \qquad (5.114)$$

More generally, at $\epsilon = 0$, $a_0 - y = F'(y)f_1(y)$ determines

$$f_1(y) = \frac{a_0 - y}{F'(y)} = \frac{1}{1 + y} \qquad (5.115)$$

while the ϵ terms imply that

$$\begin{cases} a_1 = f_1'(1)f_1(1) = -\frac{1}{8} \\ \text{and} \\ f_2(y) = \frac{a_1 - f_1'(y)f_1(y)}{F'(y)} = \frac{(1+y)^3 - 8}{8(1-y)(1+y)^4}. \end{cases} \qquad (5.116)$$

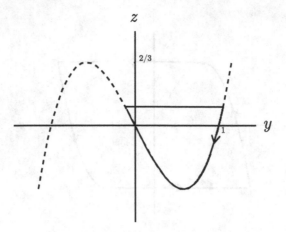

Figure 5.11: Canard sans tête

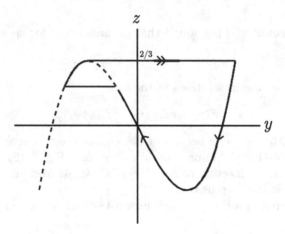

Figure 5.12: Canard avec tête

The French mathematicians used MACSYMA to get the expansions for a general f up to fifty terms while Zvonkin and Shubin [538] later provided recursion relations for them. For the van der Pol equation,

$$a = 1 - \frac{\epsilon}{8} - \frac{3}{32}\epsilon^2 + \dots \qquad (5.117)$$

and

$$z = \frac{y^3}{3} - y - \frac{\epsilon}{y+1} - \frac{\epsilon^2}{8(y+1)^2}(y^4 + 4y + 7) + \dots. \qquad (5.118)$$

These results were confirmed by Eckhaus [134] using classical (standard) analysis. Canalis-Durand [67] showed that the divergent series are of class Gevrey -1. Zvonkin and Shubin conclude

Ducks and all phenomena connected with them can be effectively discovered by numerical computations for such *moderately infinitely small* values of ϵ as $\frac{1}{10}$ or $\frac{1}{20}$.

Exercise

Find canard solutions for the van der Pol equation with $\epsilon = 1/10$ and a nearly $0.9863132\ldots$. Picture them. Imagine doing so in North Africa 35 years ago!
Beware:

A canard's life is short.

If $\bar{a}$ is a duck value, any other duck value a satisfies

$$|a - \bar{a}| = e^{-\frac{1}{k\epsilon}}$$

for some $k > 0$. Once more, we're involved with exponential asymptotics. Some, though not the originators, say the hard-to-detect phenomenon was called a canard after the French newspaper slang for a hoax. For more details, see the Scholarpedia article by Wechselberger. (Also, note Braaksma [58].)

(d) Averaging and Renormalization Group Methods

The first appendix to Sanders et al. [430] presents a brief history of the method of *averaging*. Important contributions to that approach were made by Lagrange, van der Pol, Krylov, Bogoliubov, and Mitropolsky. See Samoilenko [429] for a survey of Soviet work.

A basic underlying idea concerns *variation of parameters* (i.e., variation of constants). Its linear version is elementary and well known, while its nonlinear version has recently been attributed to Alekseev [6], rather than Lagrange or Poisson, though it seems present in the earlier celestial mechanics literature (cf. Pollard [397] and Verhulst [499]). Consider the vector initial value problem

$$\dot{z} = f(z, t, \epsilon), \quad z(t_0) = a, \tag{5.119}$$

with f depending smoothly on ϵ, and suppose the unperturbed problem

$$\dot{x} = f(x, t, 0), \quad x(t_0) = a \tag{5.120}$$

has a known solution which we will denote (as in dynamical systems) by

$$x = \phi(a, t). \tag{5.121}$$

If we seek the solution of the given problem (5.119) using the ansatz

$$z = \phi(p, t) \tag{5.122}$$

for a *variable* function $p(t)$ subject to the initial condition $p(t_0) = a$, differentiation implies that

$$\dot{z} = \frac{\partial \phi}{\partial p}\frac{dp}{dt} + \frac{\partial \phi}{\partial t} = f(\phi, t, \epsilon).$$

But $\frac{\partial \phi}{\partial t} = f(\phi, t, 0)$ and $\frac{\partial \phi}{\partial p}$ is (at least locally) invertible by the existence-uniqueness theorem. Thus, p must satisfy the initial value problem

$$\frac{dp}{dt} = \left(\frac{\partial \phi}{\partial p}\right)^{-1}(f(\phi(p,t),t,\epsilon) - f(\phi(p,t),t,0)), \quad p(t_0) = a. \qquad (5.123)$$

Because $\frac{dp}{dt} = O(\epsilon)$, p will be *slowly varying*, so it can only change substantially on a long $O(1/\epsilon)$ time interval. Solving the nonlinear problem (5.123) for $p(t)$, numerically or otherwise, determines the desired solution $z = \phi(p, t)$ of (5.119). The constant a in (5.121) could be any parameter, not just the initial value.

Adrianov et al. [9] also consider the system (5.119). They reduce it to the *standard form*

$$\dot{z} = \epsilon Z(t, z, \epsilon) \qquad (5.124)$$

by simply making a change of variables

$$z = G(t, x) \qquad (5.125)$$

where

$$\frac{\partial G}{\partial t} = f + O(\epsilon), \qquad (5.126)$$

$\frac{\partial G}{\partial x}$ is invertible, and

$$\dot{z} = \left(\frac{\partial G}{\partial x}\right)^{-1}\left(f - \frac{\partial G}{\partial t}\right) \equiv \epsilon Z(t, z, \epsilon). \qquad (5.127)$$

Examples

1. For the nearly linear vector problem

$$\dot{z} = A(t)z + \epsilon g(z, t), \quad z(0) = a, \qquad (5.128)$$

the limiting problem $\dot{x} = A(t)x$, $x(0) = a$ has the unique solution

$$x = \Phi(t)a$$

where the *fundamental matrix* $\Phi(t)$ (cf., e.g., Bellman [35]) satisfies the linear homogeneous matrix initial value problem

$$\dot{\Phi} = A(t)\Phi, \quad \Phi(0) = I.$$

(In particular, $\Phi(t)$ is the matrix exponential e^{At} when $A(t)$ is constant). Its n columns provide a full set of linearly independent solutions to (5.128) with $\epsilon = 0$, spanning all solutions. Assuming $\Phi(t)$ is available, variation of parameters determines the solution of the given problem (5.128) in the form

$$z(t) = \Phi(t)p(t). \tag{5.120}$$

Thus $\dot{z} = \dot{\Phi}p + \Phi\dot{p} = A\Phi p + \epsilon g(\Phi p, t)$ implies that the slowly varying amplitude $p(t)$ must satisfy the nonlinear vector initial value problem

$$\dot{p} = \epsilon\Phi^{-1}(t)g(\Phi(t)p, t), \quad p(0) = a. \tag{5.130}$$

Existence of $p(t)$ is guaranteed locally and its numerical solution is straight-forward.

2. In the special case of nearly linear autonomous scalar oscillators

$$\ddot{x} + x = \epsilon h(x, \dot{x}) \tag{5.131}$$

with prescribed initial values $x(0)$ and $\dot{x}(0)$, we naturally write the solution for $\epsilon = 0$ in *polar coordinates* as

$$x = r\sin(t + \phi) \quad \text{and} \quad \dot{x} = r\cos(t + \phi)$$

for constants r and ϕ determined directly by the initial values. For $\epsilon \neq 0$, we use the traditional variation of parameters approach to instead introduce variable functions $R(t)$ and $\Psi(t)$ so that

$$x = R\sin(t + \Psi) \quad \text{and} \quad \dot{x} = R\cos(t + \Psi). \tag{5.132}$$

Differentiating the first expression with respect to t and comparing that result to the second requires R and Ψ to satisfy

$$\dot{R}\sin(t + \Psi) + R\cos(t + \Psi)\dot{\Psi} = 0.$$

Likewise, differentiating the second expression implies that

$$\ddot{x} + x = \dot{R}\cos(t + \Psi) - R\sin(t + \Psi)\dot{\Psi} = \epsilon h(x, \dot{x}).$$

Using Cramer's rule, we solve these two linear equations to obtain

$$\begin{cases} \dot{R} = \epsilon\cos(t + \Psi)\, h(R\sin(t + \Psi), R\cos(t + \Psi)) \\ \text{and} \\ R\dot{\Psi} = \epsilon\sin(t + \Psi)\, h(R\sin(t + \Psi), R\cos(t + \Psi)). \end{cases} \tag{5.133}$$

The resulting nonlinear initial value problem with

$$R(0) = r \quad \text{and} \quad \Psi(0) = \phi \tag{5.134}$$

will have a unique slowly varying solution $\begin{pmatrix} R \\ \Psi \end{pmatrix}$ for bounded τ values. This variation of parameters approach is the basis of a variety of averaging procedures. For Duffing's equation, for example, $h(x, \dot{x}) = -x^3$ so

$$\begin{cases} \dot{R} = -\epsilon R^3 \cos(t + \Psi) \sin^3(t + \Psi) \\ \text{and} \\ \dot{\Psi} = -\epsilon R^2 \sin^4(t + \Psi), \end{cases}$$

a problem we could solve numerically. Several different approximate methods will be found subsequently. For more examples, see O'Malley and Kirkinis [378].

The fundamental idea to vary arbitrary constants was contained in J.-L. Lagrange's *Analytical Mechanics* [277] in 1788. A 1997 translation of the 1811 edition states:

> For problems of mechanics which can only be resolved by approximate methods, the first solution is ordinarily found by considering only the primary forces acting on the bodies. In order to extend this solution to the secondary forces which are called perturbing forces, the simplest approach is to keep the form of the first solution by considering as variables the arbitrary constants contained in it. If the quantities which were neglected and which we want to take into account, are very small, the new variables will be nearly constant and the ordinary methods of approximation could be applied. Thus the difficulty is reduced to finding the equations between these variables.

If we naively seek a solution $\begin{pmatrix} R(t, \epsilon) \\ \Psi(t, \epsilon) \end{pmatrix}$ of (5.133–5.134) as a regular power series in ϵ, the first term will be a constant, which will cause the second term to grow like t as $t \to \infty$. This is equivalent to expanding the right-hand side as a Fourier series and realizing that a nonzero constant term would result in secular behavior for later terms because that part of the forcing resonates with the complementary solution of the leading-order homogeneous system. It might again be advisable to introduce a near-identity transformation to eliminate such secularities in accordance with the Fredholm alternative. The end result is to replace the system (5.133) for R and Ψ by its *average* over its 2π period, i.e. we use the autonomous averaged equation

$$\dot{\overline{R}} = \epsilon f_1(\overline{R}) \equiv \frac{\epsilon}{2\pi} \int_0^{2\pi} \cos s \, h(\overline{R} \sin s, \overline{R} \cos s) \, ds \quad \text{with} \quad \overline{R}(0) = r. \quad (5.135)$$

This nonlinear, but separable, initial value problem provides $\overline{R}$ uniquely as a function of the slow time

$$\tau = \epsilon t,$$

i.e.

$$\tau = \int_r^{\overline{R}} \frac{ds}{f_1(s)}. \tag{5.136}$$

(Finding $\overline{R}(\tau)$ explicitly won't always be possible, though we know $\overline{R}$ will move monotonically to a zero of f_1.) Using $\overline{R}$ in place of R makes approximate sense since R should only change by a small $O(\epsilon)$ amount over a period. Then, we can directly integrate the corresponding averaged equation

$$\dot{\overline{\Psi}} = \frac{\epsilon}{\overline{R}} f_2(\overline{R}) \equiv \frac{\epsilon}{2\pi\overline{R}} \int_0^{2\pi} \sin s \, h(\overline{R}\sin s, \overline{R}\cos s) \, ds$$

to provide the approximate phase

$$\overline{\Psi}(\tau) = \phi + \int_0^\tau \frac{f_2(\overline{R}(\tau))}{\overline{R}(\tau)} \, d\tau \tag{5.137}$$

in terms of $\overline{R}$. (A more complete argument, using near-identity transformations, is given, e.g., in Rand [408].) The resulting approximation (5.132) for x and $\dot{x}$ can be shown, by a *Gronwall inequality* argument (cf. Cesari [74]), to have an $O(\epsilon)$ error on any $0 \le t \le O(1/\epsilon)$ time interval. As a final caveat, we point out that blowup later is always an *ultimate* possibility. Verhulst [500] uses the separable equation

$$\dot{x} = 2\epsilon x^2 \sin^2 t$$

as an example. With $x(0) = 1$, the exact solution is $x(t, \epsilon) = \frac{1}{1-\epsilon t + \frac{\epsilon}{2}\sin 2t}$. Moreover, when the forcing in $\dot{x} = \epsilon f(x, t, \epsilon)$ isn't periodic, one can conveniently use the *long-time average*

$$f_0(x) \equiv \lim_{T\to\infty} \frac{1}{T} \int_0^T f(x, s, 0) \, ds.$$

Higher-order averaging is also well studied and important in applications, as are extensions to even longer time intervals. An unusual variation is given in Coppola and Rand [97]. E [130] describes how the corresponding *homogenization* technique can be applied to certain elliptic and Hamilton–Jacobi equations. (See Bensoussan et al. [39], Bakhvalov and Panasenko [21], Bakhvalov et al. [22], and Holmes [209] as well.)

For the van der Pol equation, $h = (1 - x^2)\dot{x}$, so the averaged equations are easy to solve, viz.

$$\dot{\overline{R}} = \frac{\epsilon}{2}\overline{R}\left(1 - \frac{\overline{R}^2}{4}\right) \quad \text{and} \quad \dot{\overline{\Psi}} = 0.$$

This indicates that a steady-state limit cycle will occur (when $r > 0$) with $\overline{R}(\infty) = 2$.

The intimate relationship between two-timing and averaging, in general, becomes clear when one examines the asymptotic solution of the initial value problem for the vector system

$$\dot{x} = \epsilon f(x, t, \epsilon)$$

in the so-called *periodic standard form* (cf. Murdock [335], Sarlet [431], and Mudavanhu et al. [332]).

See de Jager and Jiang [224] for a variety of worked out examples of averaging.

3. The simple linear initial value problem

$$\ddot{y} + 2\epsilon\dot{y} + y = 0, \quad t \geq 0, \quad y(0) = 0, \quad \dot{y}(0) = 1 \qquad (5.138)$$

with small damping has the exact solution

$$y(t, \epsilon) = \frac{e^{-\epsilon t}}{\sqrt{1 - \epsilon^2}} \sin\left(\sqrt{1 - \epsilon^2}\, t\right). \qquad (5.139)$$

Although this solution is bounded, its regular perturbation expansion about $\epsilon = 0$ breaks down as t becomes unbounded, since secular terms arise. If we had naively set

$$y(t, \epsilon) = y_0(t) + \epsilon y_1(t) + \dots, \qquad (5.140)$$

we would need $\ddot{y}_0 + y_0 = 0$, $\ddot{y}_1 + y_1 = -2\dot{y}_0$, etc., so

$$y_0(t) = \alpha_0 \cos t + \beta_0 \sin t \qquad (5.141)$$

for constants α_0 and β_0 and $\ddot{y}_1 + y_1 = 2\alpha_0 \sin t - 2\beta_0 \cos t$. This implies the secular term

$$y_1(t) = (\alpha_1 - \alpha_0 t) \cos t + (\beta_1 - \beta_0 t) \sin t \qquad (5.142)$$

Renormalization group (or *RG*) *methods* (cf. Chen et al. [80] and [81] and Ziane [536]) proceed by eliminating the unbounded terms in a naive expansion (5.140) by replacing the ϵ-dependent initial amplitudes

$$\alpha_0 + \epsilon\alpha_1 + \dots \quad \text{and} \quad \beta_0 + \epsilon\beta_1 + \dots$$

that arise by instead using *slowly varying* amplitudes

$$A(\tau, \epsilon) \quad \text{and} \quad B(\tau, \epsilon)$$

depending on the slow time

$$\tau = \epsilon t \qquad (5.143)$$

For bounded τ, then, they could simply seek an asymptotic solution for (5.138) in the form

$$y(t, \tau, \epsilon) = A(\tau, \epsilon) \cos t + B(\tau, \epsilon) \sin t \qquad (5.144)$$

Then

$$\dot{y} = \left(B + \epsilon \frac{dA}{d\tau}\right)\cos t + \left(-A + \epsilon \frac{dB}{d\tau}\right)\sin t$$

while

$$\ddot{y} = \left(-A + 2\epsilon \frac{dB}{d\tau} + \epsilon^2 \frac{d^2 A}{d\tau^2}\right)\cos t + \left(-B - 2\epsilon \frac{dA}{d\tau} + \epsilon^2 \frac{d^2 D}{d\tau^2}\right)\sin t,$$

so (5.138) requires

$$\ddot{y} + 2\epsilon\dot{y} + y = \epsilon\left[2\frac{dB}{d\tau} + B + \epsilon\left(\frac{d^2 A}{d\tau^2} + \frac{dA}{d\tau}\right)\right]\cos t$$

$$+ \epsilon\left[-2\frac{dA}{d\tau} - A + \epsilon\left(\frac{d^2 B}{d\tau^2} + \frac{dB}{d\tau}\right)\right]\sin t = 0. \tag{5.145}$$

The linear independence of the trig functions implies that the amplitudes A and B must satisfy the initial value problem

$$2\frac{dA}{d\tau} + 2A = \epsilon\left(\frac{d^2 B}{d\tau^2} + 2\frac{dB}{d\tau}\right), \quad A(0, \epsilon) = 0$$

$$2\frac{dB}{d\tau} + 2B = -\epsilon\left(\frac{d^2 A}{d\tau^2} + 2\frac{dA}{d\tau}\right), \quad B(0, \epsilon) = 1. \tag{5.146}$$

Using series expansions

$$A(\tau, \epsilon) \sim \sum_{j \geq 0} A_j(\tau)\epsilon^j \quad \text{and} \quad B(\tau, \epsilon) \sim \sum_{j \geq 0} B_j(\tau)\epsilon^j, \tag{5.147}$$

we will need $\frac{dA_0}{d\tau} + A_0 = 0$, $A_0(0) = 0$ and $\frac{dB_0}{d\tau} + B_0 = 0$, $B_0(0) = 1$, so

$$A_0(\tau) = 0 \quad \text{and} \quad B_0(\tau) = e^{-\tau} \tag{5.148}$$

completely specify the limiting solution $y_0 = e^{-\tau}\sin t$. Next, we will need

$$2\frac{dA_1}{d\tau} + 2A_1 = \frac{d^2 B_0}{d\tau^2} + 2\frac{dB_0}{d\tau} = -e^{-\tau}, \quad A_1(0) = 0$$

and

$$2\frac{dB_1}{d\tau} + 2B_1 = -\frac{d^2 A_0}{d\tau^2} - 2\frac{dA_0}{d\tau} = 0, \quad B_1(0) = 0,$$

so

$$A_1(\tau) = -\frac{\tau}{2}e^{-\tau} \quad \text{and} \quad B_1(\tau) = 0. \tag{5.149}$$

Thus, we have the renormalized solution

$$y(t, \tau, \epsilon) = e^{-\tau}\sin t - \frac{\epsilon\tau}{2}e^{-\tau}\cos t + O(\epsilon^2) \tag{5.150}$$

for τ finite. For improved approximations, we must let the frequency of the trig functions vary with ϵ^2. For details, see O'Malley and Kirkinis [377]. (Recall that the exact solution (5.139) is a function of the fast time $\sqrt{1 - \epsilon^2}t$ and the slow time $\tau = \epsilon t$. Kirkinis [252] provides an alternative elimination technique using *cumulants* (cf. Small [464]) to describe renormalization that is somewhat closer to the original ideas of Goldenfeld, Oono, and coworkers.

4. The linear initial value problem

$$\ddot{x} + x = \epsilon \sin t, \quad t \geq 0, \quad x(0) = 1, \quad \dot{x}(0) = 0 \qquad (5.151)$$

can be solved exactly using variation of parameters. Thus, $x(t, \epsilon) = \cos t + \frac{\epsilon}{2}(\sin t - t \cos t)$. A more insightful representation, however, is to write $x(t, \tau, \epsilon) = \left(1 - \frac{\tau}{2}\right) \cos t + \frac{\epsilon}{2} \sin t$ for $\tau = \epsilon t$. This suggests that we directly seek an asymptotic solution of the form

$$x(t, \tau, \epsilon) = \alpha(\tau, \epsilon) \cos t + \epsilon \beta(\tau, \epsilon) \sin t \qquad (5.152)$$

for bounded τ values and undetermined slowly varying coefficients α and β. Then

$$\dot{x} = \epsilon \frac{d\alpha}{d\tau} \cos t - \alpha \sin t + \epsilon^2 \frac{d\beta}{d\tau} \sin t + \epsilon \beta \cos t$$

and

$$\ddot{x} = \epsilon^2 \frac{d^2\alpha}{d\tau^2} \cos t - 2\epsilon \frac{d\alpha}{d\tau} \sin t - \alpha \cos t + \epsilon^3 \frac{d^2\beta}{d\tau^2} \sin t + 2\epsilon^2 \frac{d\beta}{d\tau} \cos t - \epsilon \beta \sin t.$$

Substituting into (5.151), we obtain

$$\epsilon^2 \left(\frac{d^2\alpha}{d\tau^2} + 2\frac{d\beta}{d\tau} \right) \cos t + \epsilon \left(-2\frac{d\alpha}{d\tau} + \epsilon^2 \frac{d^2\beta}{d\tau^2} - 1 \right) \sin t = 0$$

with

$$\alpha(0, \epsilon) = 1 \quad \text{and} \quad \frac{d\alpha}{d\tau}(0, \epsilon) + \beta(0, \epsilon) = 0.$$

The linear independence of $\sin t$ and $\cos t$ implies that α and β must satisfy the initial value problem

$$\begin{cases} 2\frac{d\alpha}{d\tau} + 1 = \epsilon^2 \frac{d^2\beta}{d\tau^2}, \quad \alpha(0, \epsilon) = 1 \\ \text{and} \\ 2\frac{d\beta}{d\tau} + \frac{d^2\alpha}{d\tau^2} = 0, \quad \beta(0, \epsilon) + \frac{d\alpha}{d\tau}(0, \epsilon) = 0. \end{cases} \qquad (5.153)$$

We will solve the problem asymptotically for finite τ using power series

$$\begin{pmatrix} \alpha(\tau, \epsilon) \\ \beta(\tau, \epsilon) \end{pmatrix} \sim \sum_{j \geq 0} \begin{pmatrix} \alpha_j(\tau) \\ \beta_j(\tau) \end{pmatrix} \epsilon^j. \qquad (5.154)$$

The leading terms require

$$2\frac{d\alpha_0}{d\tau} + 1 = 0, \quad \alpha_0(0) = 1$$

and

$$2\frac{d\beta_0}{d\tau} + \frac{d^2\alpha_0}{d\tau^2} = 0, \quad \beta_0(0) + \frac{d\alpha_0}{d\tau}(0) = 0.$$

Thus

$$\alpha_0(\tau) = 1 - \frac{\tau}{2} \quad \text{and} \quad \beta_0(\tau) = \frac{1}{2} \tag{5.155}$$

No further corrections are needed!

Exercises

1. Use two-timing to show that the asymptotic solution to the initial value problem

$$\ddot{y} + \epsilon\dot{y}|\dot{y}| + y = 0, \quad y(0) = 1, \quad \dot{y}(0) = 0$$

satisfies

$$y(t,\tau) = \frac{\cos t}{1 + \frac{4\tau}{3\pi}} + O(\epsilon)$$

for $\tau = \epsilon t$ bounded.

Hint: The $\sin t$ coefficient in the Fourier expansion of $\sin t |\sin t|$ is $\frac{8}{3\pi}$ (cf. Mattheij et al. [309]).

2. (cf. Smith [466]). Consider the initial value problem for a weakly coupled electrical circuit

$$\begin{cases} (1 - \epsilon^2)Q_1'' + Q_1 = \epsilon Q_2, & Q_1(0) = q, \quad Q_1'(0) = 0 \\ (1 - \epsilon^2)Q_2'' + Q_2 = \epsilon Q_1, & Q_2(0) = 0 = Q_2'(0) \end{cases}$$

and determine a solution of the form

$$\begin{aligned} Q_1(t) &= \epsilon A(\tau, \epsilon) \sin t + B(\tau, \epsilon) \cos t \\ Q_2(t) &= C(\tau, \epsilon) \sin t + \epsilon D(\tau, \epsilon) \cos t \end{aligned}$$

for $0 \le t \le \frac{1}{\epsilon}$ using averaging. Note the *beat* phenomenon!

3. (cf. O'Malley and Kirkinis [378]). Seek a solution of the Riccati equation

$$\dot{x} = -x^2 + \epsilon x$$

of the form

$$x(t,\epsilon) = \frac{A}{1 + At}$$

and show that the slowly varying coefficient A is given by

$$A(t,\epsilon) = \frac{\epsilon x(0)}{x(0)(1 - \epsilon t - e^{-\epsilon t}) + \epsilon}.$$

Historical Remarks

The very important work of Krylov, Bogoliubov, and Mitropolsky on averaging and its applications was somewhat delayed in reaching the West from Kiev. The 1937 Russian monograph by Krylov and Bogoliubov didn't appear in English until an abridged translation by Solomon Lefschetz was published by Princeton University Press [267] in 1947. The 1955 monograph by Bogoliubov and Mitropolsky, *Asymptotic Methods in the Theory of Nonlinear Oscillations*, appeared in an English version [51] by Hindustan Publishing of New Delhi in 1961 (distributed in the West by Gordon and Breach). (Its first sixty pages are now available on Google Scholar.) For many years, two-timing was justified (cf. Morrison [327] and Perko [391]) because it gave the same results as averaging. Now, direct proofs are known (cf., e.g., Murdock [336] and Murdock and Wang [337]).

The physicists Chen, Goldenfeld, and Oono [81] presented their *renormalization group* method as a unified approach to global asymptotic analysis. They, indeed, succeeded in finding asymptotic approximations for solutions to a wide variety of challenging problems from the literature. Figuring out what their fundamental ideas are is not simple, but their aim to provide a universal technique is one all can subscribe to. Their underlying technique is to eliminate secular terms in a regular perturbation expansion by replacing constants by slowly varying amplitudes (or *envelopes*) that satisfy appropriate RG equations to be integrated as initial value problems (cf. also Kunihiro [270] and [271], Ei et al. [139], and Kirkinis [253]). See Goldenfeld [172] for an ambitious update. It is clear that the renormalization group approach is a *resummation* technique, closely related to averaging methods that suppress secular terms. The recent work of DeVille et al. [118] and Roberts [418] emphasizes that it often produces an asymptotic solution in the classical Poincaré-Birkhoff *normal form* (cf. Guckenheimer and Holmes [188], Sanders et al. [430], and Nayfeh [344]). Woodruff [528, 529] independently developed a related *invariance* method (which deserves more attention than it has received), applying it to systems of the form

$$\dot{x} = M(\epsilon^\alpha t)x + \epsilon N(\epsilon^\alpha t, t, x)$$

for $\alpha = 1$ or 2, where the matrix M has distinct, nonzero, purely imaginary eigenvalues. Cheng [82] presents a hybrid scheme combining renormalization and two-timing while Chiba [86, 87] provides a simplified RG method.

In section 3.7A of Oono [381], a proto-renormalization group approach is developed for autonomous differential equations

$$Ly = \epsilon N(y)$$

where L is a constant-coefficient linear differential operator and N is nonlinear. A somewhat analogous procedure is developed in Mudavanhu and O'Malley [331].

In the following chapter, we shall give the renormalization group method a new and somewhat simplified presentation, which we hope will be further developed as a unification of many methods found in the literature.

Chapter 6

A Simple Multi-scale Procedure for Both Oscillatory and Boundary Layer Problems

(a) Examples

In this final chapter, we shall develop a simple multiscale method by applying it to a varied sequence of examples. Readers can consult O'Malley and Kirkinis [377] for additional details.

1. A Linear Equation with Small Damping

Reconsider the linear equation

$$\ddot{y} + 2\epsilon\dot{y} + y = 0 \quad \text{on } t \geq 0 \tag{6.1}$$

with

$$y(0) = 0 \quad \text{and} \quad \dot{y}(0) = 1$$

prescribed for a small damping coefficient $\epsilon > 0$. Applying the regular perturbation procedure

$$y(t, \epsilon) \sim \sum_{k \geq 0} y_k(t)\epsilon^k,$$

© Springer International Publishing Switzerland 2014   183
R. O'Malley, *Historical Developments in Singular Perturbations*,
DOI 10.1007/978-3-319-11924-3_6

we obtain secular terms y_k for $k \geq 1$ that grow like $t^k e^{\pm it}$ as $t \to \infty$ (cf. the last example of the preceding chapter). To get an asymptotic solution valid on the moderately long time interval with

$$\tau = \epsilon t$$

finite, we have to modify our naive regular perturbation approach (which still remains valid for finite t) and instead seek an asymptotic solution in the two-time form

$$y(t, \tau, \epsilon) = \mathcal{A}(\tau, \epsilon)e^{it} + \overline{\mathcal{A}(\tau, \epsilon)}e^{-it} = \mathcal{A}(\tau, \epsilon)^{it} + \text{c.c.} \qquad (6.2)$$

where $\mathcal{A}$ is a slowly varying complex-valued amplitude to be determined for finite τ and where c.c. represents the complex conjugate function. Since the chain rule implies that $\dot{y} = \left(i\mathcal{A} + \epsilon\frac{d\mathcal{A}}{d\tau}\right)e^{it} + \text{c.c.}$ and $\ddot{y} = \left(\mathcal{A} + 2i\epsilon\frac{d\mathcal{A}}{d\tau} + \epsilon^2\frac{d^2\mathcal{A}}{d\tau^2}\right)e^{it} + \text{c.c.}$,

$$\ddot{y} + 2\epsilon\dot{y} + y = \epsilon\left(2i\frac{d\mathcal{A}}{d\tau} + 2i\mathcal{A} + \epsilon\frac{d^2\mathcal{A}}{d\tau^2} + 2\epsilon\frac{d\mathcal{A}}{d\tau}\right)e^{it} + \text{c.c.} = 0 \qquad (6.3)$$

requires $\mathcal{A}$ to satisfy the regularly perturbed *amplitude equation*

$$\frac{d\mathcal{A}}{d\tau} + \mathcal{A} = \frac{\epsilon i}{2}\left(\frac{d^2\mathcal{A}}{d\tau^2} + 2\frac{d\mathcal{A}}{d\tau}\right). \qquad (6.4)$$

Moreover, the representation (6.2) and the initial conditions

$$y(0) = \mathcal{A}(0, \epsilon) + \overline{\mathcal{A}}(0, \epsilon) = 0$$

and

$$\dot{y}(0) = i\left(\mathcal{A}(0, \epsilon) - \overline{\mathcal{A}}(0, \epsilon)\right) + \epsilon\left(\frac{d\mathcal{A}}{d\tau}(0, \epsilon) + \frac{d\overline{\mathcal{A}}}{d\tau}(0, \epsilon)\right) = 1 \qquad (6.5)$$

specify the real and imaginary parts of $\mathcal{A}(0, \epsilon)$ termwise when we take

$$\mathcal{A}(\tau, \epsilon) = \mathcal{A}_0(\tau) + \epsilon\mathcal{A}_1(\tau) + \dots \qquad (6.6)$$

as a power series in ϵ. Thus, $A(0, \epsilon) = -\frac{i}{2} - \epsilon\,\text{Re}\left(\frac{d\mathcal{A}}{d\tau}(0, \epsilon)\right)$. Substituting (6.6) into (6.4) and (6.5) yields $\frac{d\mathcal{A}_0}{d\tau} + \mathcal{A}_0 = 0$, so

$$\mathcal{A}_0(\tau) = e^{-\tau}\mathcal{A}_0(0) \quad \text{for} \quad \mathcal{A}_0(0) = -\frac{i}{2} \qquad (6.7)$$

while $\frac{d\mathcal{A}_1}{d\tau} + \mathcal{A}_1 = \frac{i}{2}\left(\frac{d^2\mathcal{A}_0}{d\tau^2} + 2\frac{d\mathcal{A}_0}{d\tau}\right)$ provides

$$\mathcal{A}_1(\tau) = -\frac{i}{2}\mathcal{A}_0(0)\tau e^{-\tau}, \qquad (6.8)$$

etc. (since $A_1(0) = -\text{Re}\left(\frac{d\mathcal{A}_0}{d\tau}(0)\right) = 0$.)

These results are satisfactory for finite τ values, but to get an asymptotic solution on an even longer time interval, we can introduce the even slower time

$$\kappa = \epsilon^2 t \tag{6.9}$$

and set

$$\mathcal{A}(\tau, \epsilon) \equiv e^{-\tau} \mathcal{B}(\kappa, \epsilon^2) \tag{6.10}$$

for the *rescaled* amplitude $\mathcal{B}$. Substituting (6.10) into (6.4), $\mathcal{B}$ must satisfy

$$\frac{d\mathcal{B}}{d\kappa} + \frac{i}{2}\mathcal{B} = \frac{\epsilon^2 i}{2}\frac{d^2\mathcal{B}}{d\kappa^2} \tag{6.11}$$

Since $\mathcal{B}(\kappa, 0) = e^{-\frac{i\kappa}{2}}\mathcal{A}_0(0)$, we obtain the leading term approximation

$$
\begin{aligned}
y(t, \tau, \kappa, \epsilon) &\sim -\frac{i}{2}e^{it}e^{-\epsilon t}e^{-\frac{i\epsilon^2 t}{2}} + \ldots + \text{c.c.} \\
&\sim -\frac{i}{2}e^{(i-\epsilon-\frac{i\epsilon^2}{2}+\ldots)t} + \ldots + \text{c.c..}
\end{aligned}
\tag{6.12}
$$

This agrees asymptotically with the exact solution

$$y(t, \epsilon) = \frac{e^{-\epsilon t}}{\sqrt{1 - \epsilon^2}} \sin\left(\sqrt{1 - \epsilon^2}\,t\right)$$

for finite κ and also corresponds to using a *multi-scale ansatz*

$$y(t, \epsilon) = e^{-\epsilon t}e^{i(1+\epsilon^2\sigma_1+\epsilon^4\sigma_2+\ldots)t}\mathcal{C}(\epsilon^2) + \text{c.c.} \tag{6.13}$$

for initially unspecified constants $\sigma_1, \sigma_2, \ldots$ (cf. Rubenfeld [428]). Alternatively, we could directly renormalize by introducing slowly varying coefficients depending on both $\tau = \epsilon t$ and $\kappa = \epsilon^2 t$ to again get a valid approximation for finite κ. Tradeoffs between the order of approximation attained and the asymptotic size of the time interval of validity can also be made (cf. Murdock [335]). Henceforth, we shall avoid introducing further slow times, like κ, but readers will realize that using such multivariable expansions can be both valuable and necessary, depending on the degree of approximation desired and the time interval of interest.

We note that Eckhaus [135] describes how the Ginzburg–Landau equation arises as the amplitude (or envelope) equation for a broad class of differential equations. Moloney and Newell [325] and Ostrovsky and Potapov [382] include applications to optics. Also see Schmid and Henningson [440]. Haberman [190] labels the method of multiple scales the method of *slow variation*. This reflects the independent development of the multiple scales technique by D. J. Benney and his MIT students (including M. Ablowitz, R. Haberman, C. Lange, and A. Newell, among others). (Benney himself was a student of the late C.-C. Lin, a prominent student of von Kármán in 1944). Related methods for chemical oscillations can be found in Kuramoto [272].

2. Duffing's Equation

Let's next reconsider Duffing's equation

$$\ddot{y} + y + \epsilon y^3 = 0, \quad t \geq 0 \tag{6.14}$$

with

$$y(0) = 1 \quad \text{and} \quad \dot{y}(0) = 0. \tag{6.15}$$

A regular perturbation (or *naive*) expansion $\Sigma y_\kappa(t)\epsilon^\kappa$ provides the series solution

$$
\begin{aligned}
y(t,\epsilon) = {} & \alpha(\epsilon)e^{it} + \epsilon\left(-\frac{3}{2}i\alpha|\alpha|^2 te^{it} + \frac{1}{8}\alpha^3 e^{3it}\right) \\
& + \epsilon^2\left(-\frac{9}{8}\alpha|\alpha|^4 t^2 e^{it} - \frac{15}{16}i\alpha|\alpha|^4 te^{it} + \frac{9}{16}i\alpha^3|\alpha|^2 te^{3it}\right. \\
& \left. - \frac{21}{64}\alpha^3|\alpha|^2 e^{3it} + \frac{1}{64}\alpha^5 e^{5it}\right) + \ldots + \text{c.c.}.
\end{aligned}
\tag{6.16}
$$

Here, the complex constant $\alpha(\epsilon)$ can be uniquely determined asymptotically as a power series in ϵ from applying the prescribed initial values. As briefly introduced in Chap. 5, the renormalization group (or RG) method characteristically drops all the unbounded terms (as $t \to \infty$) in the naive expansion and replaces the amplitude $\alpha(\epsilon)$ in the remaining expansion by a slowly varying function $A(\tau, \epsilon)$ of the slow-time $\tau = \epsilon t$ that remains to be determined (cf. Chen et al. [81]). Thus, for Duffing's equation, it seeks an asymptotic solution in the appealing form

$$
\begin{aligned}
y(t,\tau,\epsilon) = {} & A(\tau,\epsilon)e^{it} + \frac{\epsilon}{8}A^3(\tau,\epsilon)e^{3it} + \frac{\epsilon^2}{64}\left(-21A^3(\tau,\epsilon)|A(\tau,\epsilon)|^2 e^{3it}\right. \\
& \left. + A^5(\tau,\epsilon)e^{5it}\right) + \ldots + \text{c.c.}.
\end{aligned}
\tag{6.17}
$$

Substituting (6.17) back into equation (6.14) requires A to satisfy the *amplitude* equation

$$\frac{dA}{d\tau} = A|A|^2\left(\frac{3i}{2} - \frac{15}{16}\epsilon|A|^2 + \frac{i\epsilon^2}{128}|A|^4 + \ldots\right) \tag{6.18}$$

for finite $\tau \geq 0$. Note that the coefficients in (6.18) are precisely the coefficients of the corresponding secular terms te^{it} in the naive expansion (6.16). We now take

$$A(0,\epsilon) = \alpha(\epsilon).$$

Thus, the amplitude expansion

$$A(\tau,\epsilon) \sim \Sigma A_j(\tau)\epsilon^j$$

is readily determined termwise from the initial value problem for (6.18) on intervals where τ remains bounded. This RG procedure works, although it is somewhat inefficient since it first obtains the naive expansion and then eliminates its unbounded terms.

We shall now instead directly seek the asymptotic solution of Duffing's equation (6.14) for finite τ as the two-scale expansion

$$y(t, \tau, \epsilon) = A(\tau, \epsilon)e^{it} + \epsilon B(\tau, \epsilon)e^{3it} + \epsilon^2 C(\tau, \epsilon)e^{5it} + \ldots + \text{c.c.} \qquad (6.19)$$

for undetermined slowly varying complex coefficients A, B, C, $\ldots$ with power series expansions in ϵ. We will ultimately obtain an amplitude equation for A and formulas to obtain B, C, and later coefficients in terms of A. A somewhat analogous approach was taken by Noble and Hussain [352], who refer to their expansion as "generalized harmonic balance." That work, according to correspondence with Ben Noble, was continuing in 1990, but the author is unaware of later publications. Equation (6.19) implies that

$$\dot{y} = \left(iA + \epsilon\frac{dA}{d\tau}\right)e^{it} + \epsilon\left(3iB + \epsilon\frac{dB}{d\tau}\right)e^{3it}$$
$$+ \epsilon^2\left(5iC + \epsilon\frac{dC}{d\tau}\right)e^{5it} + \ldots + \text{c.c.},$$

$$\ddot{y} = \left(-A + 2i\epsilon\frac{dA}{d\tau} + \epsilon^2\frac{d^2A}{d\tau^2}\right)e^{it} + \epsilon\left(-9B + 6i\epsilon\frac{dB}{d\tau} + \ldots\right)e^{3it}$$
$$+ \epsilon^2\left(-25C + \ldots\right)e^{5it} + \ldots + \text{c.c.},$$

and

$$y^3 = A^3\epsilon^{3it} + 3A|A|^2 e^{it}$$
$$+ 3\epsilon(A^2 B e^{5it} + 2|A|^2 B e^{3it} + \overline{A}^2 B e^{it}) + \ldots + \text{c.c.},$$

so

$$\ddot{y} + y + \epsilon y^3 = \epsilon e^{it}\left[2i\frac{dA}{d\tau} + 3A|A|^2 + \epsilon\left(\frac{d^2A}{d\tau^2} + 3\overline{A}^2 B\right) + \ldots\right]$$
$$+ \epsilon e^{3it}\left[-8B + A^3 + 6\epsilon\left(\frac{idB}{d\tau} + |A|^2 B\right) + \ldots\right] \qquad (6.20)$$
$$+ 3\epsilon^2 e^{5it}\left[-8C + A^2 B + \ldots\right] + \ldots + \text{c.c.} = 0.$$

Linear independence of the powers e^{ijt} forces us to take

$$2i\frac{dA}{d\tau} + 3A|A|^2 + \epsilon\left(\frac{d^2A}{d\tau^2} + 3\overline{A}^2 B\right) + \ldots = 0,$$

$$-8B + A^3 + 6\epsilon\left(i\frac{dB}{d\tau} + |A|^2 B\right) + \ldots = 0,$$

and

$$-8C + A^2 B + \ldots = 0.$$

Iterating, then, we obtain

$$B = \frac{A^3}{8}\left(1 - \frac{21}{8}\epsilon|A|^2 + \ldots\right) \tag{6.21}$$

and

$$C = \frac{A^5}{64} + \ldots, \tag{6.22}$$

where A must satisfy the amplitude equation

$$\frac{dA}{d\tau} = \frac{3i}{2}A|A|^2\left(1 - \frac{5\epsilon}{8}|A|^2 + \ldots\right). \tag{6.23}$$

This is consistent with what we found more indirectly using renormalization. Getting more terms is straightforward, especially when using computer algebra.

We can easily determine A from (6.23) by using polar coordinates

$$A = Re^{i\phi}. \tag{6.24}$$

First, $R^2 = A\overline{A}$ implies that $2R\frac{dR}{d\tau} = \frac{dA}{d\tau}\overline{A} + A\frac{d\overline{A}}{d\tau} = 0(\epsilon^2)$, so, indeed,

$$R(\tau, \epsilon) = |\alpha(\epsilon)| \equiv R(0, \epsilon) \tag{6.25}$$

(as is classically found using the Poincaré–Lindstedt method of strained coordinates (cf. e.g., Nayfeh [341])). Next, $e^{i\phi} = \frac{A}{R}$ implies that $ie^{i\phi}\frac{d\phi}{d\tau} = \frac{d}{d\tau}\left(\frac{A}{R}\right) = \frac{1}{R}\frac{dA}{d\tau}$, so integrating

$$\frac{d\phi}{d\tau} = \frac{3}{2}R^2 - \frac{15}{16}\epsilon R^4 + O(\epsilon^2),$$

we get

$$\phi(\tau, \epsilon) = \arg(\alpha(\epsilon)) + \frac{3}{2}R^2\left(1 - \frac{5}{8}\epsilon R^2 + \ldots\right)\tau, \tag{6.26}$$

Roberts [418] obtains analogous expansions using normal form techniques, for which he also provides computer algebra implementations. Reexpansion of the solution (6.19) for finite t agrees with the naive expansion.

To extend the validity of our results to time scales where τ is infinite, one can introduce a frequency normalization, as proposed by Cheng [82], by replacing t in the phase by

$$s = t(1 + \beta_1\epsilon + \beta_2\epsilon^2 + \cdots)$$

for appropriate β_js. This corresponds to two-timing using the slow time $\tau = \epsilon t$ and a fast time s.

We note that an alternative expansion in terms of real trigonometric functions could also be used as a natural ansatz, i.e., we could seek a solution in the form

$$x(t, \tau, \epsilon) = \alpha(\tau, \epsilon) \cos t + \beta(\tau, \epsilon) \sin t$$
$$+ \epsilon(\gamma(\tau, \epsilon) \cos 3t + \delta(\tau, \epsilon) \sin 3t) + \dots$$

for slowly varying real coefficients α, β, γ, δ, ... to be determined termwise for finite τ. See Zhang et al. [535] for a generalization of this technique.

Exercises

1. (a) Obtain the naive expansion for the initial value problem

    $$\ddot{y} + y + \epsilon y^2 = 0, \quad y(0) = 1, \quad \dot{y}(0) = 0$$

    through $O(\epsilon^2)$ terms.

    (b) Determine the form of the asymptotic solution for $t = O(1/\epsilon^2)$ and the leading term in the solution of the amplitude equation.

2. (cf. O'Malley and Kirkinis [378]) Show that the initial value problem for the nearly linear equation

    $$\ddot{y} + 4\dot{y} + 3y = \epsilon y^2$$

    has an asymptotic solution of the form

    $$y(t, \epsilon) = \alpha(\tau, \epsilon)e^{-t} + \beta(\tau, \epsilon)e^{-3t} + \epsilon \left[\gamma(\tau, \epsilon)e^{-2t} + \delta(\tau, \epsilon)e^{-4t} + \zeta(\tau, \epsilon)e^{-6t} \right]$$
    $$+ O(\epsilon^2)$$

    where the slowly varying coefficients α, β, γ, δ, and ζ vary with $\tau = \epsilon t$.

3. (cf. Verhulst [504]) Solve the linear equation

    $$\ddot{y} + \epsilon(2 - e^{-\epsilon t})\dot{y} + y = 0$$

    by seeking a solution of the form

    $$y(t, \epsilon) = A(\tau, \epsilon) \cos t + B(\tau, \epsilon) \sin t$$

    with real coefficients A and B, depending on $\tau = \epsilon t$.

3. A Linear Equation with a Turning Point

Following Chen et al. [81], we next consider the linear oscillator

$$\ddot{y} + (1 - \epsilon t)y = 0 \tag{6.27}$$

on $t \geq 0$ with initial values

$$y(0) = 1 \quad \text{and} \quad \dot{y}(0) = 0. \tag{6.28}$$

Since (6.27) has the form

$$\ddot{y} + \omega^2(\tau)y = 0,$$

for the slowly varying coefficient

$$\omega(\tau) = \sqrt{1 - \tau} \quad \text{and} \quad \tau = \epsilon t,$$

we might expect the asymptotic solution to depend on both the fast time

$$T \equiv \frac{1}{\epsilon} \int_0^\tau \omega(s) \, ds = \frac{2}{3\epsilon}(1 - (1 - \tau)^{3/2}) \tag{6.29}$$

and the slow time τ (for motivation, cf. Bourland and Haberman [54], O'Malley and Williams [380], Ablowitz [1], and Haberman [190]). Thus, we shall seek an asymptotic solution to (6.27) of the form

$$y(T, \tau, \epsilon) = A(\tau, \epsilon)e^{iT} + \text{c.c.}. \tag{6.30}$$

Since $\ddot{y} = \left(-\omega^2 A + 2i\omega\epsilon \frac{dA}{d\tau} + i\epsilon \frac{d\omega}{d\tau} A + \epsilon^2 \frac{d^2 A}{d\tau^2}\right)e^{iT} + \text{c.c.}$, the given differential equation forces the amplitude A to satisfy

$$2\omega \frac{dA}{d\tau} + \frac{d\omega}{d\tau} A = \epsilon i \frac{d^2 A}{d\tau^2}. \tag{6.31}$$

Introducing the power series

$$A(\tau, \epsilon) = A_0(\tau) + \epsilon A_1(\tau) + \ldots, \tag{6.32}$$

we obtain

$$A_0(\tau) = \frac{A_0(0)}{(1 - \tau)^{1/4}} \tag{6.33}$$

to provide the leading-order approximation

$$y(t, \epsilon) = \frac{1}{(1 - \epsilon t)^{1/4}} \cos\left(\frac{2}{3\epsilon}(1 - (1 - \epsilon t))^{3/2}\right) + O(\epsilon) \tag{6.34}$$

which agrees with the leading asymptotic approximation of the exact solution

$$y(t, \epsilon) = \pi \left[Bi'\left(-\frac{1}{\epsilon^{2/3}}\right) Ai\left(\frac{\epsilon t - 1}{\epsilon^{2/3}}\right) - Ai'\left(-\frac{1}{\epsilon^{2/3}}\right) Bi\left(\frac{\epsilon t - 1}{\epsilon^{2/3}}\right) \right] \tag{6.35}$$

(using the Airy functions Ai and Bi, solutions of the Airy equation $z'' = xz$) and with the WKB approximation. For finite t, (6.34) also agrees with the approximation $y(t, \epsilon) \sim e^{\epsilon t/4} \cos\left(t - \frac{\epsilon}{4}t^2\right)$ of Chen et al. [81], but (6.34) also holds for $\tau < 1$, i.e. up to the turning point, a singular point of the limiting amplitude equation.

The related equation

$$\ddot{y} + e^{-\epsilon t}y = 0$$

is solved by the Bessel functions $J_0\left(\frac{2}{\epsilon}e^{-\epsilon t/2}\right)$ and $Y_0\left(\frac{2}{\epsilon}e^{-\epsilon t/2}\right)$, while its two-time expansion holds for

$$\epsilon e^{\epsilon t/2} \ll 1$$

(cf. Cheng and Wu [84]).

Ablowitz [1] shows how analogous multiple scale arguments can be used for the nonlinear pendulum equation

$$\ddot{y} + \rho^2(\epsilon t)\sin y = 0$$

while Johnson [226] considers

$$\ddot{y} + a(\epsilon t)y + b(\epsilon t)y^3 = 0,$$

i.e. Duffing's equation with slowly varying coefficients. Nayfeh [341] earlier considered the linear equation

$$\ddot{y} + p(\epsilon t, \epsilon)\dot{y} + q(\epsilon t, \epsilon)y = r(\epsilon t, \epsilon)e^{i\phi(t,\epsilon)}.$$

Likewise, the first-order initial value problem

$$\dot{y} + 2\epsilon t y = 0, \quad y(0) = 1$$

has the exact solution

$$y(t, \epsilon) = e^{-\epsilon t^2},$$

depending on the nonlinear scale $T = \epsilon t^2$.

4. The Duffing–van der Pol Equation

Benney and Newell [37] introduced the Duffing–van der Pol equation

$$\ddot{y} + y + \epsilon y^3 + \epsilon^2(y^2 - 1)\dot{y} = 0. \tag{6.36}$$

It was later treated by other authors, including Mudavanhu and O'Malley [330]. We shall naturally seek an asymptotic solution

$$y(t, \epsilon) = A(\tau, \epsilon)e^{it} + \epsilon B(\tau, \epsilon)e^{3it} + \epsilon^2 C(\tau, \epsilon)e^{5it} + \ldots + \text{c.c.} \tag{6.37}$$

for $\tau = \epsilon t$ finite. Differentiating (6.37) and collecting coefficients of e^{it}, e^{3it}, e^{5it}, ... in (6.36), we find that A must satisfy the amplitude equation

$$\frac{dA}{d\tau} = \frac{3i}{2}A|A|^2 - \frac{\epsilon A}{2}(|A|^2 - 1) + \frac{\epsilon i}{2}\frac{d^2A}{d\tau^2} + \ldots$$

while

$$B \sim \frac{A^3}{8} \quad \text{and} \quad C \sim \frac{A^2 B}{8}.$$

Iterating, we rewrite the amplitude equation as

$$\frac{dA}{d\tau} = \frac{3i}{2} A|A|^2 + \frac{\epsilon A}{2} \left[1 - |A|^2 - \frac{27i}{4}|A|^4 + \ldots \right] \tag{6.38}$$

and we will solve it using polar coordinates

$$A = R e^{i\phi} \tag{6.39}$$

to provide the separated equations

$$\frac{dR}{d\tau} = \frac{\epsilon R}{2}(1 - R^2) + \ldots \tag{6.40}$$

and

$$\frac{d\phi}{d\tau} = \frac{3}{2}R^2 \left(1 + \frac{9\epsilon}{4}R^2 + \ldots \right) \tag{6.41}$$

for the amplitude R and phase ϕ. Note that (6.40) shows that R is actually a function of the slower time

$$\kappa = \epsilon\tau = \epsilon^2 t. \tag{6.42}$$

Setting

$$R(\kappa, \epsilon) = R_0(\kappa) + \epsilon R_1(\kappa) + \ldots, \tag{6.43}$$

R_0 must satisfy the Bernoulli equation

$$\frac{dR_0}{d\kappa} = \frac{R_0}{2} \left(1 - R_0^2 \right). \tag{6.44}$$

We can explicitly obtain R_0. It decays to the steady-state 1 as $\kappa \to \infty$ (presuming $R_0(0) \neq 0$). Knowing R to any order in ϵ, we next integrate (6.41) to find ϕ asymptotically.

5.   Morrison's Counterexample

Morrison [327] introduced the oscillator

$$\ddot{y} + y + \epsilon(\dot{y}^3 + 3\epsilon y) = 0 \tag{6.45}$$

as a counterexample to traditional two-timing. Proceeding as above, we set

$$y(t, \tau, \epsilon) = A(\tau, \epsilon)e^{it} + \epsilon B(\tau, \epsilon)e^{3it} + \epsilon^2 C(\tau, \epsilon)e^{5it} + \ldots + \text{c.c.} \tag{6.46}$$

for undetermined coefficients A, B, C, ... depending on the slow variable $\tau = \epsilon t$ and ϵ. Differentiating and separating coefficients of successive odd powers of e^{it} in (6.46), we find that

$$B \sim -\frac{A^3}{8}\left(i + \frac{45}{8}\epsilon|A|^2 + \dots\right) \tag{6.47}$$

and

$$C \sim \frac{3}{64}A^5 + \dots, \tag{6.48}$$

while A must satisfy the amplitude equation

$$\frac{dA}{d\tau} + 3iA|A|^2 + \frac{3}{2}A|A|^4 + \frac{3\epsilon}{2}A\left(1 + \frac{3i}{8}|A|^4\right) + \dots = 0. \tag{6.49}$$

When we set $A = Re^{i\phi}$, we get the separated equations

$$\frac{dR}{d\tau} = -\frac{3}{2}(R^3 + \epsilon R + \dots) \tag{6.50}$$

and

$$\frac{d\phi}{d\tau} = -\frac{3}{2}R^2\left(1 + \frac{3\epsilon}{8}R^2 + \dots\right).$$

Integrating the ϕ equation is straightforward once R is known, but integrating (6.50) for $R(\tau, \epsilon)$ is a little tricky. If we balance the first three terms, we get a Bernoulli equation for $R(\kappa, \epsilon)$ whose nontrivial solutions decay to zero as

$$\kappa = \epsilon^2 t = \epsilon\tau \tag{6.51}$$

tends to infinity. The linear third term in (6.50) dominates the second when $R = O(\epsilon^{1/2})$, so ignoring it gets two-timing in trouble (cf. the apologetic discussion in Kevorkian and Cole [249] and Exercise 6 at the end of Chap. 2). Likewise, one can consider the solvable initial value problem for $\dot{y} = \epsilon y^2 - \epsilon^2 y$.

Exercise

The system

$$\begin{cases} \dot{x} = x(y - \epsilon y^2) \\ \dot{y} = -\ln x - \epsilon y \end{cases}$$

studied by Chiba [85] reduces to

$$\ddot{y} + y + \epsilon(\dot{y} - y^2) = 0$$

when $\ln x$ is eliminated from the system. Show that the appropriate ansatz to obtain the asymptotic solution is

$$y(t, \tau, \epsilon) = A(\tau, \epsilon)e^{it} + \epsilon(B(\tau, \epsilon) + C(\tau, \epsilon)e^{2it}) + \epsilon^2 D(\tau, \epsilon)e^{3it} + \dots + \text{c.c.}$$

where A, B, C, D, etc. are slowly varying functions of $\tau = \epsilon t$ for finite τ.

6.   A Homogeneous Linear Initial Value Problem

For the scalar linear equation

$$\epsilon \ddot{y} + a(t)\dot{y} + b(t)y = 0, \quad t \geq 0 \tag{6.52}$$

with initial values $y(0)$ and $\dot{y}(0)$ prescribed, presuming

$$a(t) > 0 \tag{6.53}$$

and that a and b are smooth, we can expect y to converge to the solution of the reduced problem

$$a(t)\dot{Y}_0 + b(t)Y_0 = 0, \quad Y_0(0) = y(0)$$

for bounded intervals of $t \geq 0$. To describe the nonuniform convergence of $\dot{y}$ near $t = 0$, we introduce the natural stretched variable

$$\eta = \frac{1}{\epsilon} \int_0^t a(s)\, ds \tag{6.54}$$

and the two-time ansatz

$$y(t, \eta, \epsilon) = A(t, \epsilon) + \epsilon B(t, \epsilon)e^{-\eta}. \tag{6.55}$$

Then $\dot{y} = \dot{A} + (-aB + \epsilon\dot{B})e^{-\eta}$ and $\epsilon\ddot{y} = \epsilon\ddot{A} + (a^2 B - 2\epsilon a\dot{B} - \epsilon\dot{a}B + \epsilon^2\ddot{B})e^{-\eta}$, so (6.52) implies that

$$\epsilon\ddot{A} + a\dot{A} + bA + \epsilon(-a\dot{B} - \dot{a}B + bB + \epsilon\ddot{B})e^{-\eta} = 0.$$

The linear independence of 1 and $e^{-\eta}$ imply that the outer expansion $A(t, \epsilon)$ must, as expected, satisfy

$$\epsilon\ddot{A} + a\dot{A} + bA = 0 \tag{6.56}$$

as a power series $\Sigma A_\kappa \epsilon^\kappa$ in ϵ, while $B(t, \epsilon)$ must satisfy

$$a\dot{B} + \dot{a}B - bB = \epsilon\ddot{B} \tag{6.57}$$

as a series $\Sigma B_\kappa \epsilon^\kappa$. Moreover, (6.55) converts the initial conditions to

$$\begin{cases} A(0, \epsilon) + \epsilon B(0, \epsilon) = y(0) \\ \text{and} \\ \dot{A}(0, \epsilon) - a(0)B(0, \epsilon) + \epsilon\dot{B}(0, \epsilon) = \dot{y}(0). \end{cases} \tag{6.58}$$

Since $A_0(t)$ must satisfy the initial value problem

$$a(t)\dot{A}_0 + b(t)A_0 = 0, \quad A_0(0) = y(0),$$

$$A_0(t) = e^{-\int_0^t \frac{b(s)}{a(s)} ds} \frac{a(0)}{a(t)} y(0) \tag{6.59}$$

while

$$a(t)\dot{B}_0 + (\dot{a} - b)B_0 = 0, \quad B_0(0) = (\dot{A}_0(0) - \dot{y}(0))/a(0)$$

implies that

$$B_0(t) = -e^{\int_0^t \frac{b(s)}{a(s)} ds} \frac{1}{a(t)} \left(\dot{y}(0) + \frac{b(0)y(0)}{a(0)} \right). \tag{6.60}$$

Higher-order terms in the expansions for A and B follow uniquely without complication. When the outer limit A_0 decays asymptotically as $t \to \infty$, the expansion may even remain valid for unbounded t (cf. Hoppensteadt [212]). Otherwise, we should only use the results for finite t values.

7.   A Linear Two-Point Problem

For the two-point linear problem

$$\epsilon y'' + a(x)y' + b(x)y = c(x), \quad 0 \le x \le 1 \tag{6.61}$$

with

$$a(x) > 0, \tag{6.62}$$

smooth coefficients a, b, and c, and for prescribed bounded endvalues

$$y(0) \quad \text{and} \quad y(1),$$

we will find an asymptotic solution in the form

$$y(x, \eta, \epsilon) = A(x, \epsilon) + B(x, \epsilon)e^{-\eta} \tag{6.63}$$

for the (fast or) boundary layer variable

$$\eta = \frac{1}{\epsilon} \int_0^x a(s)\, ds. \tag{6.64}$$

(Thus, as Wasow's thesis showed, we again have an initial layer.) Since

$$\epsilon y'' + a(x)y' + b(x)y = \epsilon A'' + a(x)A' + b(x)A$$
$$+ (\epsilon B'' - aB' + (b - a')B)e^{-\eta} = c(x)$$

while, up to asymptotically negligible terms,

$$A(0, \epsilon) + B(0, \epsilon) = y(0) \quad \text{and} \quad A(1, \epsilon) \sim y(1),$$

the outer solution A must satisfy the terminal value problem

$$\epsilon A'' + a(x)A' + b(x)A = c(x), \quad A(1, \epsilon) = y(1) \tag{6.65}$$

as a power series, while B must satisfy the initial value problem

$$\epsilon B'' - aB' + (b - a')B = 0, \quad B(0, \epsilon) = y(0) - A(0, \epsilon). \tag{6.66}$$

Occasionally, such problems have a trivial boundary layer correction B. An example is provided by

$$\epsilon y'' + y' + y = 1 + x, \qquad y(0) = 0 \text{ and } y(1) = 1$$

since the outer solution

$$A(x, \epsilon) = x$$

is an exact solution of the two-point problem. The initial layer would reappear when the initial value is changed to a nontrivial $y(0)$.

If $a(x)$ were negative everywhere, the asymptotic solution of (6.61) would instead have the form

$$y(x, \xi, \epsilon) = C(x, \epsilon) + D(x, \epsilon)e^{-\xi} \quad \text{for} \quad \xi = \frac{1}{\epsilon} \int_x^1 a(s) \, ds,$$

i.e. it would display the analogous terminal boundary layer. In the special case

$$\epsilon y'' + (1 + x)y' - y = 0, \tag{6.67}$$

of (6.61)–(6.62), the outer solution of (6.67) is simply

$$A(x, \epsilon) = \frac{1}{2}(1 + x)y(1), \tag{6.68}$$

while the stretched variable

$$\eta = \frac{1}{\epsilon} \int_0^x (1 + s) \, ds = \frac{x}{\epsilon}\left(1 + \frac{x}{2}\right) \tag{6.69}$$

determines an initial layer correction $B(x, \epsilon)e^{-\eta}$ with

$$B(x, 0) = \frac{(y(0) - y(1)/2)}{(1 + x)^2}. \tag{6.70}$$

The full expansion (6.63) can, of course, also be obtained from the exact solution

$$y(x, \epsilon) = \frac{1}{2}(1 + x)y(1) + (1 + x)\frac{\int_x^1 \frac{1}{(1+s)^2} e^{-\frac{1}{\epsilon}\left(s + \frac{s^2}{2}\right)} ds}{\int_0^1 \frac{1}{(1+s)^2} e^{-\frac{1}{\epsilon}\left(s + \frac{s^2}{2}\right)} ds}\left(y(0) - \frac{y(1)}{2}\right)$$

using Laplace's method.

Note: We can use the WKB method to show that (6.61) has solutions of the form (6.63). The theory of Maslov [307] shows that WKB holds even more broadly.

As pointed out by Nayfeh [341], the unpublished 1951 Caltech thesis of Gordon Latta [281] used the ansatz (6.63) to solve such linear boundary value problems.

In the special case of the Frobenius equation

$$\epsilon xy'' - y' - xy = 0, \quad 0 \leq x \leq 1 \text{ with } y(0) = y(1) = 1 \qquad (6.71)$$

(from Bender and Orszag [36]), we'd expect to have a terminal layer since

$$a(x) = -\frac{1}{x} < 0.$$

Thus, we introduce the stretched variable

$$\xi = \frac{1}{\epsilon} \int_x^1 \frac{ds}{s} = -\frac{\ln x}{\epsilon}.$$

Since $e^{-\xi} = x^{1/\epsilon}$, we shall seek an asymptotic solution of (6.71) in the form

$$y(x, \epsilon) = A(x, \epsilon) + B(x, \epsilon)x^{1/\epsilon} \qquad (6.72)$$

where A and B have power series expansions in ϵ. Clearly, $x^{1/\epsilon}$ is asymptotically negligible for any fixed $x < 1$, yet very important for x near 1. Differentiating (6.72) twice, we get

$$\epsilon xy'' - y' - xy = (\epsilon xA'' - A' - xA) + \left(\epsilon xB'' + B' - \frac{B}{x} - xB\right)x^{1/\epsilon} = 0.$$

Thus, the series for A should satisfy the initial value problem

$$\epsilon xA'' - A' - xA = 0, \quad A(0, \epsilon) = 1 \qquad (6.73)$$

as a power series in ϵ, while B should satisfy

$$\epsilon x^2 B'' + xB' - (1 + x^2)B = 0, \quad B(1, \epsilon) = 1 - A(1, \epsilon). \qquad (6.74)$$

Their leading terms are

$$A_0(x) = e^{-x^2/2} \quad \text{and} \quad B_0(x) = xe^{(x^2-1)/2}\left(1 - \frac{1}{\sqrt{e}}\right). \qquad (6.75)$$

Exercises

1. Since the asymptotic solution of

$$\epsilon y' + a(x)y = -b(x),$$

with $a(x) > 0$ can be found for $x \geq 0$ in the form

$$y(x, \epsilon) = A(x, \epsilon) + e^{-\frac{1}{\epsilon} \int_0^x a(s)\,ds}(y(0) - A(0, \epsilon)),$$

we might expect to find the asymptotic solution of the corresponding Riccati equation

$$\epsilon z' = a(x)z + b(x)z^2$$

in the form

$$z(x, \epsilon) = \frac{1}{A(x, \epsilon) + e^{-\frac{1}{\epsilon}\int_0^x a(s)\, ds}k(\epsilon)}.$$

Show this by finding the explicit solution to the Riccati equation and expanding the integrals involved asymptotically.

2. (Kevorkian and Cole [249]) For the boundary value problem

$$\epsilon y'' + y' + xy = 0, \quad y(0) = 0, \quad y(1) = \sqrt{e},$$

one can find the exact solution in terms of Airy functions by using the Sturm transformation

$$y = e^{-\frac{x}{2\epsilon}}z.$$

Show that

$$y(x, \epsilon) = A_0(x) + B_0(x)e^{-x/\epsilon} + O(\epsilon)$$

where $A_0(x) = e^{x^2/2}$ and $B_0(x) = -e^{-x^2/2}$.

8.   A Fourth-Order Two-Point Problem

The linear fourth-order equation

$$\epsilon^2 y'''' - a^2(x)y'' = f(x), \quad 0 \le x \le 1 \tag{6.76}$$

for $a(x) > 0$ and with prescribed boundary values

$$y(0), \quad y'(0), \quad y(1), \quad \text{and} \quad y'(1) \tag{6.77}$$

and $\epsilon > 0$ small might arise in elasticity. We can expect nonuniform convergence of y', but not y, at both endpoints, according to Wasow [511]. Thus, we will seek an asymptotic solution of the form

$$y(x, \epsilon) = A(x, \epsilon) + \epsilon B(x, \epsilon)e^{-\frac{1}{\epsilon}\int_0^x a(s)\, ds} + \epsilon C(x, \epsilon)e^{-\frac{1}{\epsilon}\int_x^1 a(s)\, ds} \tag{6.78}$$

with power series for A, B, and C. Then

$$y'' = A'' + \left(\epsilon B'' - 2aB' - a'B + \frac{a^2}{\epsilon}B\right)e^{-\frac{1}{\epsilon}\int_0^x a(s)\, ds}$$

$$+ \left(\epsilon C'' + 2aC' + a'C + \frac{a^2}{\epsilon}C\right)e^{-\frac{1}{\epsilon}\int_x^1 a(s)\, ds}$$

and

$$
\begin{aligned}
y'''' = A'''' + \Big[&\epsilon B'''' - 4aB''' - 6a'B'' - 4a''B' - a'''B \\
&+ \frac{1}{\epsilon}\left[6a^2B'' + 12aa'B' + 3(a')^2B + 4aa''B\right] \\
&+ \frac{1}{\epsilon^2}\left[-4a^3B' - 6a^2a'B\right] + \frac{a^4B}{\epsilon^3}\Big]e^{-\frac{1}{\epsilon}\int_0^x a(s)\,ds} \\
&+ \Big[\epsilon C'''' + 4aC''' + 6a'C'' + 4a''C' + a'''C \\
&+ \frac{1}{\epsilon}\left[6a^2C'' + 12aa'C' + 3(a')^2C + 4aa''C\right] \\
&+ \frac{1}{\epsilon^2}(4a^3C' + 6a^2a'C) + \frac{a^4C}{\epsilon^3}\Big]e^{-\frac{1}{\epsilon}\int_x^1 a(s)\,ds}.
\end{aligned}
$$

Substituting into the differential equation (6.76) and separating the coefficients of 1, $e^{-\frac{1}{\epsilon}\int_0^x a(s)\,ds}$, and $e^{-\frac{1}{\epsilon}\int_x^1 a(s)\,ds}$ requires us to impose the decoupled linear equations

$$
\epsilon^2 A'''' - a^2 A'' = f, \tag{6.79}
$$

$$
\begin{aligned}
\epsilon^3 B'''' - \epsilon^2(4aB''' + 6a'B'' + 4a''B' + a'''B) + \epsilon(5a^2B'' + 12aa'B' \\
+ 3(a')^2B + 4aa''B) - (2a^3B' - 5a^2a'B) = 0,
\end{aligned} \tag{6.80}
$$

and

$$
\begin{aligned}
\epsilon^3 C'''' + \epsilon^2(4aC''' + 6a'C'' + 4a''C' + a'''C) + \epsilon(5a^2C'' + 12aa'C' \\
+ 3(a')^2C + 4aa''C) + (2a^3C' + 5a^2a'C) = 0
\end{aligned} \tag{6.81}
$$

for power series A, B, and C. Moreover, the boundary conditions imply that

$$
\begin{cases}
y(0) \sim A(0, \epsilon) + \epsilon B(0, \epsilon) \\
y'(0) \sim A'(0, \epsilon) - a(0)B(0, \epsilon) + \epsilon B'(0, \epsilon) \\
y(1) \sim A(1, \epsilon) + \epsilon C(1, \epsilon) \\
\text{and} \\
y'(1) \sim A'(1, \epsilon) + a(1)C(1, \epsilon) + \epsilon C'(1, \epsilon).
\end{cases} \tag{6.82}
$$

As a result, the outer expansion $A(x, \epsilon)$ must satisfy the two-point outer problem

$$
\epsilon^2 A'''' - a^2 A'' = f, \quad A(0, \epsilon) \sim y(0) - \epsilon B(0, \epsilon), \\
\text{and } A(1, \epsilon) \sim y(1) - \epsilon C(1, \epsilon), \tag{6.83}
$$

the initial layer's amplitude $B(x, \epsilon)$ will satisfy

$$
-2aB' + 5a'B = O(\epsilon), \quad a(0)B(0, \epsilon) \sim A'(0, \epsilon) - y'(0) + \epsilon B'(0, \epsilon), \tag{6.84}
$$

while $C(x, \epsilon)$ must satisfy

$$2aC' + 5a'C = O(\epsilon), \quad a(1)C(1, \epsilon) \sim y'(1) - A'(1, \epsilon) - \epsilon C'(1, \epsilon). \quad (6.85)$$

Thus, A_0 follows as a solution of the two-point problem

$$A_0'' = \frac{f(x)}{a^2(x)} \text{ on } 0 \le x \le 1 \text{ with } A_0(0) = y(0) \text{ and } A_0(1) = y(1) \quad (6.86)$$

(i.e., a cancellation law applies), while B_0 and C_0 must satisfy

$$\frac{B_0'}{B_0} = \frac{5a'}{2a} \text{ on } 0 \le x \le 1 \text{ with } a(0)B_0(0) = -y'(0) + A_0'(0), \quad (6.87)$$

and

$$\frac{C_0'}{C_0} = -\frac{5a'}{2a} \text{ on } 0 \le x \le 1 \text{ with } a(1)C_0(1) = y'(1) - A_0'(1). \quad (6.88)$$

Specifically,

$$A_0(x) = y(0) + \left(y(1) - y(0) - \int_0^1 (1-s)\frac{f(s)}{a^2(s)} \, ds \right) x + \int_0^x (x-s)\frac{f(s)}{a^2(s)} \, ds,$$

$$B_0(x) = \left(\frac{a(x)}{a(0)} \right)^{5/2} B_0(0)$$

and

$$C_0(x) = \left(\frac{a(1)}{a(x)} \right)^{5/2} C_0(1).$$

Later terms also follow without complication.

9. A Linear Two-Point Problem with a Turning Point

The linear equation

$$\epsilon y'' - x(x+2)y' + xy = 0, \quad -1 \le x \le 1 \quad (6.89)$$

was considered by Hemker [202]. We will use boundary values

$$y(-1) = 1 \quad \text{and} \quad y(1) = 2. \quad (6.90)$$

Because the coefficient of y' is positive at $x = -1$ and negative at $x = 1$, we might expect to have $O(\epsilon)$-thick boundary layers at each endpoint.

The differential equation (6.89) has linearly independent solutions

$$y_1(x) = x + 2$$

and
$$y_2(x, \epsilon) = (x+2) \int_{-1}^{x} \frac{e^{-\frac{1}{\epsilon} \int_t^1 s(s+2)\, ds}}{(t+2)^2}\, dt, \tag{6.91}$$

so all solutions have the form $y(x, \epsilon) = (x+2)\alpha + y_2(x, \epsilon)\beta$ for ϵ-dependent constants α and β. Note that using $y_1(-1) - 1$ eliminates the need for an initial layer. Applying the boundary conditions, we get the exact solution

$$y(x, \epsilon) = x + 2 - \frac{y_2(x, \epsilon)}{y_2(1, \epsilon)} \tag{6.92}$$

which features a terminal layer since

$$y_2 \sim \frac{1}{x(x+2)^2} e^{-\frac{1}{\epsilon} \int_x^1 s(s+2)\, ds}.$$

Alternatively, we might directly write the asymptotic solution of (6.89)–(6.90) as

$$y(x, \epsilon) = A(x, \epsilon) + B(x, \epsilon)e^{-\xi} \tag{6.93}$$

for

$$\xi = \frac{1}{\epsilon} \int_x^1 s(s+2)\, ds = \frac{1}{3\epsilon}(1-x)(2+x)^2. \tag{6.94}$$

Substituting (6.93) into (6.89) and separating coefficients of 1 and $e^{-\xi}$, we will need

$$\epsilon A'' - x(x+2)A' + xA = 0, \quad A(-1, \epsilon) = 1$$

and

$$\epsilon B'' + x(x+2)B' + (3x+2)B = 0, \quad B(1, \epsilon) = 2 - A(1, \epsilon). \tag{6.95}$$

Thus, as anticipated,

$$A(x, \epsilon) = x + 2 \tag{6.96}$$

and $B(x, 0)$ satisfies

$$x(x+2)B_0' + (3x+2)B_0 = 0, \quad B_0(0) = -1.$$

The singularity of B_0 at $x = 0$ is irrelevant because $B_0 e^{-\xi}$ is asymptotically negligible there.

10.   Nonlinear Two-Point Problem I

Johnson [226] considered the quasilinear two-point problem

$$\epsilon y'' + (1+x)^2 y' - y^2 = 0, \quad 0 \le x \le 1, \quad y(0) = y(1) = 2. \tag{6.97}$$

We'd naturally expect an initial layer corresponding to the stretched variable

$$\eta = \frac{1}{\epsilon} \int_0^x (1+s)^2\, ds = \frac{x}{\epsilon}\left(1 + x + \frac{x^2}{3}\right), \tag{6.98}$$

so we somewhat boldly seek an asymptotic solution of the form

$$y(x, \eta, \epsilon) = A(x, \epsilon) + B(x, \epsilon)e^{-\eta} + \epsilon C(x, \epsilon)e^{-2\eta} + \ldots \qquad (6.99)$$

Since this implies that

$$
\begin{aligned}
\epsilon y'' + (1+x)^2 y' - y^2 = {} & \epsilon A'' + (1+x)^2 A' - A^2 \\
& + [\epsilon B'' - (1+x)^2 B' - 2(1+x+A)B]e^{-\eta} \\
& + \big[\epsilon^2 C'' + \epsilon\big(-3(1+x)^2 C' - 4(1+x)C - 2AC\big) \\
& + 2(1+x)^2 C - B^2\big]e^{-2\eta} + \ldots = 0,
\end{aligned}
$$

we ask the outer solution $A(x, \epsilon)$ to satisfy the nonlinear outer problem

$$\epsilon A'' + (1+x)^2 A' - A^2 = 0, \quad A(1, \epsilon) = y(1) = 2,$$

so we simply obtain

$$A(x, \epsilon) = 1 + x. \qquad (6.100)$$

Then, $B(x, \epsilon)$ must satisfy the linear problem

$$(1+x)^2 B' + 4(1+x)B = \epsilon B'', \quad B(0, \epsilon) = -1 - \epsilon C(0, \epsilon) \qquad (6.101)$$

while $C(x, \epsilon)$ must satisfy

$$2(1+x)^2 C - B^2 - 3\epsilon((1+x)^2 C' + 2(1+x)C) + \epsilon^2 C'' = 0. \qquad (6.102)$$

Letting $B(x, \epsilon) \sim \Sigma B_j(x)\epsilon^j$ and $C(x, \epsilon) \sim \Sigma C_j(x)\epsilon^j$, we determine the coefficients termwise starting with

$$(1+x)B_0' + 4B_0 = 0, \quad B_0(0) = 1,$$
$$(1+x)^2 B_1' + 4(1+x)B_1 = B_0'', \quad B_1(0) = -C_0(0),$$

and

$$2(1+x)^2 C_0 = B_0^2$$

to obtain the uniform approximation

$$y(x, \eta, \epsilon) = 1 + x + \frac{e^{-\eta}}{(1+x)^4}$$
$$+ \frac{\epsilon e^{-\eta}}{6(1+x)^{10}}\left[37(1+x)^6 - 40(1+x)^3 + 3e^{-\eta}\right] + O(\epsilon^2).$$

$$(6.103)$$

11. Nonlinear Two-Point Problem II

Johnson [226] also considered the two-point problem

$$\epsilon y'' - \frac{y'}{1+2x} - \frac{1}{y} = 0, \quad 0 \leq x \leq 1, \quad y(0) = y(1) = 3. \tag{6.104}$$

Again, using the ansatz

$$y(x, \eta, \epsilon) = A(x, \epsilon) + B(x, \epsilon)e^{-\xi} + \epsilon C(x, \epsilon)e^{-2\xi} + \dots \tag{6.105}$$

with

$$\xi = \frac{1}{\epsilon} \int_x^1 \frac{ds}{1+2s} = -\frac{1}{2\epsilon} \ln\left(\frac{1+2x}{3}\right), \tag{6.106}$$

we separate the resulting coefficients of 1, $e^{-\xi} = \left(\frac{1+2x}{3}\right)^{1/2\epsilon}$, and $e^{-2\xi} = \left(\frac{1+2x}{3}\right)^{1/\epsilon}$ in (6.104) to get

$$\epsilon A'' - \frac{A'}{1+2x} - \frac{1}{A} = 0, \quad A(0, \epsilon) = 3, \tag{6.107}$$

$$\epsilon B'' - \frac{B'}{1+2x} - \left(\frac{1}{A^2} + \frac{2}{1+2x}\right)B = 0, \quad B(1, \epsilon) = 3 - A(1, \epsilon) - \epsilon C(1, \epsilon), \tag{6.108}$$

and

$$\epsilon^2 C'' + \epsilon \left(\frac{3C'}{1+2x} - \frac{4C}{(1+2x)^2}\right) + \frac{2C}{(1+2x)^2} = \frac{B^2}{A^3} \tag{6.109}$$

since

$$\frac{1}{y} = \frac{1}{A} - \frac{B}{A^2}e^{-\xi} + \frac{B^2 - \epsilon AC}{A^3}e^{-2\xi} + \dots$$

provided $A \neq 0$. The leading coefficients are determined by

$$\begin{cases} A_0 A_0' + (1+2x) = 0, & A_0(0) = 3, \\ \frac{A_1'}{1+2x} - \frac{A_1}{A_0^2} = A_0'', & A_1(0) = 0, \\ B_0' + \left(\frac{1+2x}{A_0^2} + 2\right)B_0 = 0, & B_0(1) = 3 - A_0(1), \\ \frac{B_1'}{1+2x} + \left(\frac{1}{A_0^2} + \frac{2}{1+2x}\right)B_1 \\ \quad = B_0'' + \frac{B_0 A_1}{A_0^3}, & B_1(1) = 3 - A_1(1) - C_0(1), \\ \text{and} \\ C_0 = \frac{(1+2x)^2 B_0^2}{2A_0^3}. \end{cases} \tag{6.110}$$

No complications occur since $A_0(x) = \sqrt{9 - 2x - 2x^2} > 0$.

12.   Nonlinear Two-Point Problem III

The related example

$$\epsilon y'' + y' + e^{-y} = 0, \quad 0 \le x \le 1 \tag{6.111}$$

using the boundary values

$$y(0) = y(1) = 0, \tag{6.112}$$

is similar to one considered by Bender and Orszag [36], Chen et al. [81], and Cheng [83]

If one used a regular perturbation expansion, one would encounter secular terms at order $O(\epsilon)$. One could renormalize to eliminate them by introducing slowly varying amplitudes (cf. O'Malley and Kirkinis [375, 376]). Correspondingly, we will attempt to find an asymptotic solution

$$y\left(x, \frac{x}{\epsilon}, \epsilon\right) = A(x, \epsilon) + B(x, \epsilon)e^{-x/\epsilon} + \epsilon C(x, \epsilon)e^{-2x/\epsilon} + O(\epsilon^2), \tag{6.113}$$

with undetermined amplitudes A, B, and C and the stretched variable

$$\eta = x/\epsilon.$$

Since $e^{-(A+Be^{-\eta})} = e^{-A}(1 - Be^{-\eta} + \frac{1}{2}B^2 e^{-2\eta} + \ldots)$, substituting (6.113) into (6.111) implies that

$$(\epsilon A'' + A' + e^{-A}) + (\epsilon B'' - B' - e^{-A}B)e^{-x/\epsilon}$$
$$+ \left(\epsilon^2 C'' - 3\epsilon C' - \epsilon e^{-A}C + 2C + \frac{1}{2}e^{-A}B^2\right)e^{-2x/\epsilon} + \ldots = 0,$$

so we ask the outer solution $A(x, \epsilon)$ to satisfy the nonlinear terminal value problem

$$\epsilon A'' + A' + e^{-A} = 0, \quad A(1, \epsilon) = 0 \tag{6.114}$$

as a power series in ϵ. Likewise, we ask $B(x, \epsilon)$ to satisfy the linear initial value problem

$$\epsilon B'' - B' - e^{-A}B = 0, \quad B(0, \epsilon) = -A(0, \epsilon) \tag{6.115}$$

while $C(x, \epsilon)$ must satisfy

$$2C + \frac{1}{2}e^{-A}B^2 - \epsilon(3C' + e^{-A}C) + \epsilon^2 C'' = 0. \tag{6.116}$$

Since $A_0' + e^{-A_0} = 0$, $A_0(1) = 0$, the limiting outer solution is

$$A_0(x) = \ln(2 - x). \tag{6.117}$$

Moreover, $B_0' + e^{-A_0} B_0 = 0$ and $B_0(0) = -\ln 2$ imply that

$$B_0(x) = \frac{1}{2}(2 - x)\ln 2 \qquad (6.118)$$

and

$$C_0(x) = -\frac{1}{4}e^{-A}B_0^2 = (x - 2)\left(\frac{\ln 2}{2}\right)^2. \qquad (6.119)$$

As the earlier phase-plane analysis suggests, some two-point problems have additional asymptotic solutions, not necessarily in the form we first envisioned. Thus, McLeod and Sadhu [311] and Bakri et al. [23] both find an initially unbounded solution to

$$\epsilon y'' + 2y' + e^y = 0, \qquad y(0) = y(1) = 0,$$

in addition to the asymptotic solution with an initial layer.

13. A Quasilinear Two-Point Problem

For the general quasilinear equation

$$\epsilon y'' + a(x)y' + f(x, y) = 0, \quad 0 \le x \le 1 \qquad (6.120)$$

with

$$a(x) > 0,$$

smooth functions a and f, and prescribed boundary values

$$y'(0) \quad \text{and} \quad y(1), \qquad (6.121)$$

we must expect the limiting solution $A_0(x)$ to satisfy the reduced problem

$$a(x)A_0' + f(x, A_0) = 0, \quad A_0(1) = y(1) \qquad (6.122)$$

(presuming existence of A_0 throughout $0 \le x \le 1$) and y' to have an initial layer. Thus, we will seek an asymptotic solution of the form

$$y(x, \eta, \epsilon) = A(x, \epsilon) + \epsilon B(x, \epsilon)e^{-\eta} + O(\epsilon^2 e^{-2\eta}) \qquad (6.123)$$

using the stretched variable

$$\eta = \frac{1}{\epsilon}\int_0^x a(s)\, ds. \qquad (6.124)$$

Then (6.120) implies that

$$(\epsilon A'' + (\epsilon^2 B'' - 2\epsilon a B' - \epsilon a' B + a^2 B)e^{-\eta} + \dots)$$
$$+ a(x)(A' + (\epsilon B' - aB)e^{-\eta} + \dots)$$
$$+ f(x, A + \epsilon B e^{-\eta} + \dots) = 0.$$

Expanding $f(x, A + \epsilon Be^{-\eta} + \ldots)$ about $y = A$ and separating coefficients of 1 and $e^{-\eta}$, we ask the outer expansion A to satisfy

$$\epsilon A'' + a(x)A' + f(x, A) = 0, \quad A(1, \epsilon) = y(1) \tag{6.125}$$

as a power series in ϵ. Likewise, we ask B to satisfy

$$\epsilon B'' - aB' - a'B + f_y(x, A)B = 0,$$
$$-a(0)B(0, \epsilon) + A'(0, \epsilon) + \epsilon B'(0, \epsilon) = y'(0). \tag{6.126}$$

All proceeds in a straightforward manner when $A_0(x)$ is defined throughout $0 \leq x \leq 1$. In particular, $B_0(x)$ will then satisfy the linear initial value problem

$$a(x)B_0' + a'(x)B_0 = f_y(x, A_0)B_0, \quad a(0)B_0(0) = -\frac{f(0, A_0(0))}{a(0)} - y'(0). \tag{6.127}$$

14.   A Nearly Linear Oscillator with Slowly Varying Coefficients

Consider the initial value problem for the nearly linear oscillator

$$\ddot{y} + \omega^2(\tau)y + \epsilon\kappa(\tau)y^2 = 0 \tag{6.128}$$

with $\omega > 0$ and with ω and κ being given smooth functions of the slow time

$$\tau = \epsilon t.$$

As in (6.27), we introduce the fast time

$$T = \frac{1}{\epsilon} \int_0^\tau \omega(s)\, ds \tag{6.129}$$

(replacing t) and let the solution of (6.128) follow using the two-time ansatz

$$y(T, \tau, \epsilon) = A(\tau, \epsilon)e^{iT} + \epsilon\left(B(\tau, \epsilon) + C(\tau, \epsilon)e^{2iT}\right) + O(\epsilon^2) + \text{c.c.} \tag{6.130}$$

for undetermined amplitudes $A, B, C, \ldots$. Then, the differential equation (6.128) implies that

$$\ddot{y} + \omega^2 y + \epsilon\kappa y^2 = \epsilon\left(2i\omega\frac{dA}{d\tau} + iA\frac{d\omega}{d\tau} + \epsilon\frac{d^2A}{d\tau^2} + \ldots\right)e^{iT}$$
$$+ \epsilon(\omega^2 B + 2\kappa|A|^2 + \ldots)$$
$$+ \epsilon\left(-3\omega^2 C + \kappa A^2 + 2i\epsilon\frac{d\omega}{d\tau}C + \epsilon^2\frac{d^2C}{d\tau^2} + \ldots\right)e^{2iT}$$
$$+ \ldots + \text{c.c.} = 0. \tag{6.131}$$

Thus, the amplitude A must satisfy

$$2\omega\frac{dA}{d\tau} + A\frac{d\omega}{d\tau} - i\epsilon\frac{d^2A}{d\tau^2} + \ldots = 0 \qquad (6.132)$$

while

$$\omega^2 B + 2\kappa|A|^2 + \ldots = 0 \qquad (6.133)$$

and

$$-3\omega^2 C + \kappa A^2 + \ldots = 0. \qquad (6.134)$$

Further, the initial values

$$y(0) \sim A(0,\epsilon) + \epsilon(B(0,\epsilon) + C(0,\epsilon)) + \ldots + \text{c.c.}$$

and

$$\dot{y}(0) \sim i\omega A(0,\epsilon) + \epsilon\left(\frac{dA}{d\tau}(0,\epsilon) + 2i\omega C(0,\epsilon)\right) + \ldots + \text{c.c.}$$

can be used to specify the complex initial value $A(0,\epsilon)$ termwise. The leading coefficient is

$$A_0(\tau) = \sqrt{\frac{\omega(0)}{\omega(\tau)}}A_0(0), \qquad (6.135)$$

analogous to the WKB approximation.

15. A Semilinear Two-Point Problem

Consider the semilinear scalar equation

$$\epsilon^2 y'' + f(x,y,\epsilon) = 0, \quad 0 \le x \le 1 \qquad (6.136)$$

with f smooth and with Neumann boundary conditions

$$y'(0) \quad \text{and} \quad y'(1) \quad \text{prescribed.} \qquad (6.137)$$

We will suppose the reduced problem

$$f(x, A_0, 0) = 0 \qquad (6.138)$$

has an isolated solution $A_0(x)$ defined throughout $0 \le x \le 1$ such that the resulting Jacobian matrix

$$f_y(x, A_0(x), 0) \quad \text{is strictly stable} \qquad (6.139)$$

there. Then, the implicit function theorem allows us to termwise generate an outer expansion

$$A(x,\epsilon) \sim \Sigma_{j\ge 0} A_j(x)\epsilon^j \qquad (6.140)$$

of the given equation. Thus,

$$f_y(x, A_0, 0)A_1 + f_\epsilon(x, A_0, 0) = 0$$

and

$$f_y(x, A_0, 0)A_2 + \frac{1}{2}f_{yy}(x, A_0, 0)A_1^2 + \frac{1}{2}f_{\epsilon\epsilon}(x, A_0, 0) + A_0'' = 0$$

uniquely specify A_1, A_2, and $A(x, \epsilon)$ termwise in terms of A_0. Since y' must generally have endpoint layers, we seek the asymptotic solution of (6.136)–(6.137) in the form

$$
\begin{aligned}
y(x, \epsilon) = {}& A(x, \epsilon) + \epsilon B(x, \epsilon)e^{-\frac{1}{\epsilon}\int_0^x \sqrt{-f_y(s, A(s,\epsilon),\epsilon)}\, ds} \\
& + \epsilon C(x, \epsilon)e^{-\frac{1}{\epsilon}\int_x^1 \sqrt{-f_y(s, A(s,\epsilon),\epsilon)}\, ds} + \dots
\end{aligned}
\tag{6.141}
$$

Substituting into (6.136) shows that $B(x, \epsilon)$ must satisfy

$$
\begin{cases}
2B'\sqrt{-f_y(x, A, \epsilon)} + B(\sqrt{-f_y(x, A, \epsilon)})_x = \epsilon B'' + \dots \\
B(0, \epsilon)\sqrt{-f_y(0, A(0, \epsilon), \epsilon)} = A'(0, \epsilon) - y'(0) - \epsilon B'(0, \epsilon) + \dots
\end{cases}
\tag{6.142}
$$

while

$$
\begin{cases}
2C'\sqrt{-f_y(x, A, \epsilon)} + C(\sqrt{-f_y(x, A, \epsilon)})_x = -\epsilon C'' \dots \\
C(1, \epsilon)\sqrt{-f_y(1, A(1, \epsilon), \epsilon)} = y'(1) - A'(1, \epsilon) - \epsilon C'(1, \epsilon) + \dots
\end{cases}
\tag{6.143}
$$

Thus

$$
\begin{cases}
B_0(x) = \sqrt[4]{\dfrac{f_y(0, A_0(0), 0)}{f_y(x, A_0(x), 0)}}\, B_0(0) \\
\text{and} \\
C_0(x) = \sqrt[4]{\dfrac{f_y(1, A_0(1), 0)}{f_y(x, A_0(x), 0)}}\, C_0(1),
\end{cases}
\tag{6.144}
$$

as expected.

16.   Linear Systems

Lastly, consider the linear vector initial value problem for the singularly perturbed (or slow/fast) system

$$
\begin{cases}
\dot{x} = A_{11}(t)x + A_{12}(t)y \\
\epsilon \dot{y} = A_{21}(t)x + A_{22}(t)y
\end{cases}
\tag{6.145}
$$

on a bounded interval $0 \le t \le T$ where

$$A_{22}(t) \quad \text{is a strictly stable matrix.} \tag{6.146}$$

For prescribed initial vectors $x(0)$ and $y(0)$, we shall seek the asymptotic solution

$$
\begin{cases}
x(t, \epsilon) = \alpha(t, \epsilon) + \epsilon e^{\frac{1}{\epsilon}\int_0^t A_{22}(s)\, ds}\gamma(t, \epsilon) \\
y(t, \epsilon) = \beta(t, \epsilon) + e^{\frac{1}{\epsilon}\int_0^t A_{22}(s)\, ds}\delta(t, \epsilon),
\end{cases}
\tag{6.147}
$$

presuming smooth coefficients in (6.145). We naturally expect the outer solution

$$\begin{pmatrix} \alpha(t,\epsilon) \\ \beta(t,\epsilon) \end{pmatrix} \tag{6.148}$$

to satisfy the system (6.145) as a power series

$$\Sigma_{j\geq 0} \begin{pmatrix} \alpha_j(t) \\ \beta_j(t) \end{pmatrix} \epsilon^j$$

in ϵ. In particular,

$$\begin{cases} \dot{\alpha}_0 = A_{11}\alpha_0 + A_{12}\beta_0 \\ 0 = A_{21}\alpha_0 + A_{22}\beta_0 \end{cases}$$

implies that

$$\beta_0(t) = -A_{22}^{-1}A_{21}\alpha_0, \tag{6.149}$$

leaving $\alpha_0(t)$ to satisfy the lower-dimensional initial value problem

$$\dot{\alpha}_0 = (A_{11} - A_{12}A_{22}^{-1}A_{21})\alpha_0, \quad \alpha_0(0) = x(0). \tag{6.150}$$

Imposing the initial condition

$$x(0) = \alpha(0,\epsilon) + \epsilon\gamma(0,\epsilon) \tag{6.151}$$

for the slow variable x termwise will force higher-order terms $\begin{pmatrix} \alpha_j(t) \\ \beta_j(t) \end{pmatrix}$ for $j > 0$ in the outer expansion to satisfy corresponding nonhomogeneous differential equations with the initial value

$$\alpha_j(0) = -\gamma_{j-1}(0) \tag{6.152}$$

determined by preceding terms in the initial layer correction.

Linearity requires this correction

$$e^{\frac{1}{\epsilon}\int_0^t A_{22}(s)\,ds} \begin{pmatrix} \epsilon\gamma \\ \delta \end{pmatrix} \tag{6.153}$$

to also satisfy (6.145), so

$$e^{\frac{1}{\epsilon}\int_0^t A_{22}(s)\,ds}\left(A_{22}\gamma + \epsilon\frac{d\gamma}{dt}\right) = e^{\frac{1}{\epsilon}\int_0^t A_{22}(s)\,ds}(\epsilon A_{11}\gamma + A_{12}\delta)$$

and

$$e^{\frac{1}{\epsilon}\int_0^t A_{22}(s)\,ds}\left(A_{22}\delta + \epsilon\frac{d\delta}{dt}\right) = e^{\frac{1}{\epsilon}\int_0^t A_{22}(s)\,ds}(\epsilon A_{21}\gamma + A_{22}\delta)$$

i.e.

$$(A_{22} - \epsilon A_{11})\gamma = A_{12}\delta - \epsilon\frac{d\gamma}{dt} \quad \text{and} \quad \frac{d\delta}{dt} = A_{21}\gamma.$$

Thus, we will take

$$\gamma = (A_{22} - \epsilon A_{11})^{-1} \left(A_{12}\delta - \epsilon \frac{d\gamma}{dt} \right) \tag{6.154}$$

while $\delta(t)$ must termwise satisfy

$$\frac{d\delta}{dt} = A_{21}(A_{22} - \epsilon A_{11})^{-1} A_{12}\delta - \epsilon A_{21}(A_{22} - \epsilon A_{11})^{-1}\frac{d\gamma}{dt}. \tag{6.155}$$

together with the remaining initial condition

$$\delta(0, \epsilon) = y(0) - \beta(0, \epsilon) \tag{6.156}$$

for the fast-variable y. When we introduce power series expansions

$$\begin{pmatrix} \gamma \\ \delta \end{pmatrix} \sim \Sigma_{j\geq 0} \begin{pmatrix} \gamma_j \\ \delta_j \end{pmatrix} \epsilon^j, \tag{6.157}$$

all decouples most efficiently. We get

$$\gamma_0 = A_{22}^{-1} A_{12}\delta_0 \tag{6.158}$$

where the exponential δ_0 must satisfy the initial value problem

$$\frac{d\delta_0}{dt} = A_{21}A_{22}^{-1}A_{12}\delta_0, \quad \delta_0(0) = y(0) + A_{22}^{-1}(0)A_{12}(0)x(0). \tag{6.159}$$

Higher-order coefficients satisfy corresponding nonhomogeneous problems. The process used for this example is easier than that of Tikhonov–Levinson.

Among many possible generalizations (cf. O'Malley [364]), one could analogously solve initial value problems for linear systems with two small parameters like

$$\begin{cases} \dot{x} = A_{11}(t)x + A_{12}(t)y + A_{13}(t)z + b_1(t) \\ \epsilon\dot{y} = A_{21}(t)x + A_{22}(t)y + b_2(t) \\ \mu\dot{z} = A_{31}(t)x + A_{33}(t)z + b_3(t) \end{cases}$$

on intervals $0 \leq t \leq 1$ where the matrices

$$A_{22}(t) \quad \text{and} \quad - A_{33}(t)$$

are both strictly stable, subject to prescribed boundary values

$$x(0), \ y(0), \ \text{and} \ z(1)$$

for $\mu = \epsilon$ or ϵ^2. One could also consider *kinematically similar* systems where

$$u(t) = K(t, \epsilon, \mu) \begin{pmatrix} y(t) \\ z(t) \end{pmatrix}$$

for a smooth invertible matrix K.

Note that the asymptotic forms of solutions used for the preceding examples may be anticipated by considering simpler problems with known asymptotics. Thus, we have a method to successively improve our intuition.

Exercises (cf. Zauderer [531] and O'Malley and Kirkinis [377])

1. Determine the asymptotic behavior of the solution to the initial value problem

$$\epsilon u' + (x-1)^2 u = 1, \quad u(0) = 0.$$

2. Consider the Cauchy problem

$$\epsilon(u_{tt} - c^2 u_{xx}) + u_t + u_x = 0$$

with

$$u(x,0) \quad \text{and} \quad u_t(x,0) \quad \text{prescribed for all } x.$$

Seek a asymptotic solution of the form

$$u(x,t,\epsilon) = A(x,t,\epsilon) + \epsilon B(x,t,\epsilon)e^{-t/\epsilon}.$$

3. (cf. Chen and O'Malley [79]) The boundary-value problem

$$\begin{cases} \epsilon y'' - b(x)y' - g(x,y) = 0, & 0 \le x \le 1 \\ \epsilon y'(0) = y(0) - \alpha, & y'(1) = \beta \end{cases}$$

with $b(x) > 0$ arises in chemical reactor theory. Find an asymptotic approximation of the form

$$y(x,\epsilon) = A_0(x) + \epsilon\left(A_1(x) + B_1(x)e^{-\frac{1}{\epsilon}\int_0^x b(s)\,ds}\right) + O(\epsilon^2)$$

presuming $A_0(x)$ is defined for $0 \le x \le 1$.

Note that Miller [318] and Johnson [226] find multiscale solutions for a variety of boundary value problems for partial differential equations (cf. also Kirkinis and O'Malley [254]).

(b) Exponential Asymptotics for Two-point Problems

Howls [219] introduces *transseries* or templates to obtain exact solutions for certain singular perturbation problems. In particular, he uses an infinity of exponential scales as prefactors of divergent power series. We shall simply illustrate his ideas on the linear example

$$\begin{cases} \epsilon u'' + (2x+1)u' + 2u = 0, & 0 \le x \le 1 \\ u(0) = \alpha, u(1) = \beta. \end{cases} \tag{6.160}$$

He begins with the WKB ansatz

$$u(x, \epsilon) = A(x, \epsilon) + B(x, \epsilon) e^{-F(x)/\epsilon} \sim \sum_{n=0}^{\infty} a_r(x) \epsilon^r + e^{-\frac{F(x)}{\epsilon}} \sum_{r=0}^{\infty} b_r(x) \epsilon^r \quad (6.161)$$

(as we have been using) for

$$F(x) = \int_0^x (2s + 1) \, ds = x^2 + x.$$

At $x = 1$, we make an asymptotically negligible error, i.e. $O(e^{-F(1)/\epsilon})$ when we take

$$A(1, \epsilon) \sim \sum_{r=0}^{\infty} a_r(1) \epsilon^r \sim \beta.$$

We can compensate it by adding an additional series to the solution, prefactored by $e^{-F(1)/\epsilon}$, thereby cancelling the error at $x = 1$, but it will imply that the initial conditions will then be violated asymptotically negligibly. We could next add a new series, prefactored by $e^{-(F(x)-F(1))/\epsilon}$ but that, in turn, makes an error in satisfying the terminal condition, unless we add a fifth series with prefactor $e^{-2F(1)/\epsilon}$.

To exactly satisfy the boundary conditions, Howls uses a *ladder* of series, viz.

$$u_{\text{trans}}(x, \epsilon) \sim \sum_{p=0}^{\infty} e^{-pF(1)/\epsilon} \sum_{r=0}^{\infty} a_r^p(x) \epsilon^r$$

$$+ e^{-F(x)/\epsilon} \sum_{p=0}^{\infty} e^{-pF(1)/\epsilon} \sum_{r=0}^{\infty} b_r^{(p)}(x) \epsilon^r. \quad (6.162)$$

The differential equation requires that

$$a_r^{(p)}(x) = \frac{1}{2x + 1} \left[a_{r-1,x}^{(p)}(1) - 3b_r^{(p-1)}(1) - a_{r-1,x}^{(p)}(x) \right]$$

and

$$b_{r,x}^{(p)}(x) = \frac{1}{2x - 1} b_{r-1,xx}^{(p)}(x)$$

subject to the boundary conditions

$$a_r^{(p)}(0) + b_r^{(p)}(0) = \delta_{r0} \delta_{p0} \alpha$$

and

$$a_r^{(p)}(1) + b_r^{(p-1)}(1) = \delta_{r0} \delta_{p0} \beta.$$

All coefficients are thereby determined. Using just $O(1)$ terms, he gets

$$
u_{\text{trans}}(x, \epsilon) = \left\{ \frac{3\beta}{2x+1} - \frac{(\alpha - 3\beta)}{2x+1} \sum_{p=1}^{\infty} 3^p e^{-pF(1)/\epsilon} \right.
$$

$$
\left. + e^{-\frac{F(x)}{\epsilon}} (\alpha - 3\beta) \sum_{p=0}^{\infty} 3^p e^{-pF(1)/\epsilon} \right\} (1 + O(\epsilon)).
$$

For $3e^{-F(1)/\epsilon} < 1$, the p sums converge, so

$$
u_{\text{trans}}(x, \epsilon) = \left\{ \frac{3}{2x+1} \left(\frac{\beta - \alpha e^{-F(1)/\epsilon}}{1 - 3e^{-F(1)/\epsilon}} \right) \right.
$$

$$
\left. + e^{-\frac{F(x)}{\epsilon}} \left(\frac{\alpha - 3\beta}{1 - 3e^{-F(1)/\epsilon}} \right) \right\} (1 + O(\epsilon)).
\tag{6.163}
$$

This approximation satisfies the differential equation and both boundary conditions (to lowest order). It can be carried out to higher-orders and can be related to a multiple scales expansion. The computational results hold for quite moderate values of ϵ.

In analogous fashion, we can solve the two-point problem for the linear equation

$$
\epsilon u'' + c(x)u' + d(x)u = 0
$$

using the fundamental set of WKB solutions. Extensions to nonlinear problems remain difficult, but very much worthy of study. One might, for example, consider asymptotically negligible corrections to the solutions of the nonlinear Examples 10–12.

(c) Epilog

In summary, we have demonstrated asymptotic methods to solve a wide variety of singular perturbation problems. There, however, still remains no clear-cut universal technique. Approximating solutions through asymptotic and numerical techniques remains a challenge, despite all the advances made since Prandtl and Poincaré. Yet, significant progress is certainly occurring steadily. Our study of many examples certainly suggests the appropriate ansatz for asymptotic solutions to a wide variety of problems. We hope this presentation improves your understanding and motivates and facilitates continued research and applications regarding singular perturbations.

References

[1] M.J. Ablowitz, *Nonlinear Dispersive Waves: Asymptotic Analysis and Solitons* (Cambridge University Press, Cambridge, 2011)

[2] M. Abramowitz, I.A. Stegun (eds.), *Handbook of Mathematical Functions*, vol. 55 of *National Bureau of Standards Applied Mathematics Series* (U.S. Government Printing Office, Washington, DC, 1964). Reprinted by Dover, New York

[3] R.C. Ackerberg, R.E. O'Malley, Jr., Boundary layer problems exhibiting resonance. Stud. Appl. Math. **49**, 277–295 (1970)

[4] S.B. Ai, Multi-bump solutions to Carrier's problem. J. Math. Anal. Appl. **277**, 405–422 (2003)

[5] R.C. Aiken (ed.), *Stiff Computation* (Oxford University Press, Oxford, 1985)

[6] V.M. Alekseev, An estimate for the solutions of ordinary differential equations. Vestn. Moskov. Univ. Ser. I Mat. Mech. **2**, 28–36 (1961)

[7] C.M. Andersen, J.F. Geer, Power series solutions for the frequency and period of the limit cycle of the van der Pol equation. SIAM J. Appl. Math. **42**, 678–693 (1982)

[8] J.D. Anderson, Jr., *A History of Aeronautics and Its Impact on Flying Machines* (Cambridge University Press, Cambridge, 1997)

[9] I. Andrianov, J. Awrejcewicz, L.I. Manevitch, *Asymptotical Mechanics of Thin-Walled Structures* (Springer, Berlin, 2004)

[10] I.V. Andrianov, L.I. Manevitch, *Asymptotology: Ideas, Methods, and Applications* (Kluwer, Dordrecht, 2002)

[11] D.V. Anosov, On limit cycles in systems of differential equations with a small parameter in the highest derivatives. Am. Math. Soc. Trans. **33**(2), 233–276 (1963)

© Springer International Publishing Switzerland 2014 215
R. O'Malley, *Historical Developments in Singular Perturbations*,
DOI 10.1007/978-3-319-11924-3

[12] V.I. Arnold, V.V. Kozlov, A.I. Neishtadt, *Mathematical Aspects of Classical and Celestial Mechanics*, 2nd edn. (Springer, Berlin, 1997)

[13] Z. Artstein, I.G. Kevrekides, M. Slemrod, E.S. Titi, Slow observables of singularly perturbed differential equations. Nonlinearity **20**, 2463–2481 (2007)

[14] Z. Artstein, J. Linshiz, E.S. Titi, Young measure approach to computing slowly-advancing fast oscillations. Multiscale Model Sim. **6**, 1085–1097 (2007)

[15] U.M. Ascher, L.R. Petzold, *Computer Methods for Ordinary Differential Equations and Differential-Algebraic Equations* (SIAM, Philadelphia, 1998)

[16] U.M. Ascher, R.M.M. Mattheij, R.D. Russell, *Numerical Solution of Boundary Value Problems for Ordinary Differential Equations* (Prentice-Hall, Englewood Cliffs, NJ, 1988)

[17] K.E. Avrachenkov, J.A. Filar, P.G. Howlett, *Analytic Perturbation Theory and Its Applications* (SIAM, Philadelphia, 2013)

[18] J. Awrejcewicz, V.A. Krysko, *Introduction to Asymptotic Methods* (Chapman and Hall/CRC, Boca Raton, FL, 2006)

[19] S.M. Baer, T. Erneux, J. Rinzel, The slow passage through a Hopf bifurcation: Delay, memory effects, and resonance. SIAM J. Appl. Math **49**, 55–71 (1989)

[20] B. Baillaud, H. Bourget (eds.), *Correspondance d'Hermite et de Stieltjes* (Gauthier-Villars, Paris, 1905)

[21] N.S. Bakhvalov, G.P. Panasenko, *Homogenisation: Averaging Processes in Periodic Media: Mathematical Problems in the Mechanics of Composite Materials* (Kluwer, Dordrecht, 1989)

[22] N.S. Bakhvalov, G.P. Panasenko, A.L. Štaras, The averaging method for partial differential equations (homogenization) and its applications, in *Partial Differential Equations V*, ed. by Yu.V. Egorov, M.A. Shubin (Springer, Berlin, 1999), pp. 211–247

[23] T. Bakri, Y.A. Kutnetsov, F. Verhulst, E. Doedel, Multiple solutions of a generalized singularly perturbed Bratu problem. Int. J. Bifurcation Chaos **22**(1250095), (2012)

[24] P. Ball. *Serving the Reich: The Struggle for the Soul of Physics Under Hitler* (The Bodley Head, London, 2013)

[25] W. Balser, *Formal Power Series and Linear Systems of Meromorphic Differential Equations* (Springer, New York, 2000)

[26] E.J. Barbeau, P.J. Leah, Euler's 1760 paper on divergent series. Historia Math. **3**, 141–160 (1976)

[27] G.I. Barenblatt, *Scaling, Self-Similarity, and Intermediate Asymptotics* (Cambridge University Press, Cambridge, 1996)

[28] G.I. Barenblatt, *Scaling* (Cambridge University Press, Cambridge, 2003)

[29] J. Barrow-Green, *Poincaré and the Three Body Problem*, vol. 11 of *History of Mathematics* (Amer. Math. Soc, Providence, RI, 1997)

[30] J. Barrow-Green, D. Fenster, J. Schwermer, R. Siegmund-Schultze, Emigration of mathematicians and transmission of mathematics: Historical lessons and consequences of the Third Reich. Oberwolfach Rep. **8**, 2891–2961 (2011)

[31] G. Batchelor, *The Life and Legacy of G. I. Taylor* (Cambridge University Press, Cambridge, 1994)

[32] H. Baumgärtel, *Analytic Perturbation Theory for Matrices and Operators* (Birkhäuser, Basel, 1985)

[33] R. Beals, R. Wong, *Special Functions: A Graduate Text* (Cambridge University Press, Cambridge, 2010)

[34] A. Beckenbach, Alice Beckenbach, in *Fascinating Mathematical People: Interviews and Memoirs*, ed. by D.J. Albers, G.L. Alexanderson (Princeton University Press, Princeton, NJ, 2011), p. 105

[35] R. Bellman, *Introduction to Matrix Analysis*, 2nd edn. (McGraw-Hill, New York, 1970)

[36] C.M. Bender, S.A. Orszag, *Advanced Mathematical Methods for Scientists and Engineers* (McGraw-Hill, New York, 1978)

[37] D.J. Benney, A.C. Newell, Sequential time closures for interacting random waves. J. Math. and Phys. **46**, 363–393 (1967)

[38] E. Benoit, J.L. Callot, F. Diener, M. Diener, Chasse au canard. Collect. Math. **31**(1–3), 37–119 (1981)

[39] A. Bensoussan, J.-L. Lions, G. Papanicolaou, *Asymptotic Methods in Periodic Media* (North-Holland, Amsterdam, 1978)

[40] I.P. van den Berg, *Nonstandard Asymptotic Analysis*, vol. 1249 of *Lecture Notes in Math.* (Springer, Heidelberg, 1987)

[41] N. Berglund, B. Gentz, C. Kuehn, Hunting French ducks in a noisy environment. J. Differ. Equ. **252**, 4786–4841 (2010)

[42] B. Bergmann, M. Epple, R. Unger (eds.), *Transcending Tradition: Jewish Mathematicians in German-Speaking Academic Culture* (Springer, Berlin, 2012)

[43] A. Beyerchen, *Scientists Under Hitler: Politics and the Physics Community in the Third Reich* (Yale University Press, New Haven, CT, 1977)

[44] G.D. Birkhoff, Fifty years of American mathematics. *Semicentennial Addresses of the American Mathematical Society*, vol. 2 (Amer. Math. Soc., New York, 1938), pp. 270–315

[45] N. Bleistein, R.A. Handelsman, *Asymptotic Expansions of Integrals* (Holt, Rinehart and Winston, New York, 1975)

[46] D. Bloor, Skepticism and the social construction of science and technology: The case of the boundary layer, in *The Skeptics*, ed. by S. Luper (Ashgate, Aldershot, 2003), pp. 249–265

[47] D. Bloor, *The Enigma of the Aerofoil* (University of Chicago Press, Chicago, 2011)

[48] G. Bluman et al., Julian D. Cole. Not. Am. Math. Soc. **47**(4), 466–473 (2000)

[49] E. Bodenschatz, M. Eckert, Prandtl and the Göttingen school, in *A Voyage Through Turbulence*, ed. by A. Davidson, Y. Kaneda, H.K. Moffatt, K.R. Sreenivasan (Cambridge University Press, Cambridge, 2011), pp. 40–100

[50] V.N. Bogaevski, A. Povzner, *Algebraic Methods in Nonlinear Perturbation Theory* (Springer, New York, 1991)

[51] N.N. Bogoliubov, Yu. A. Mitropolsky, *Asymptotic Methods in the Theory of Nonlinear Oscillations* (Gordon and Breach, New York, 1961)

[52] E. Borel, *Lectures on Divergent Series*, 2nd edn. (U.S. Energy Research and Development Administration, Washington, DC, 1975). Translated by C. L. Critchfield and A. Vakar

[53] D.L. Bosley, A technique for the numerical verification of asymptotic expansions. SIAM Rev. **38**, 128–135 (1996)

[54] F.J. Bourland, R. Haberman, The modulated phase shift for strongly nonlinear, slowly-varying, and weakly-damped oscillators. SIAM J. Appl. Math. **48**, 737–748 (1988)

[55] J.P. Boyd, *Weakly Nonlocal Solitary Waves and Beyond-All-Orders Asymptotics* (Kluwer, Dordrecht, 1998)

[56] J.P. Boyd, The devil's invention: Asymptotic, superasymptotic, and hyperasymptotic series. Acta Appl. Math. **56**, 1–98 (1999)

[57] J.P. Boyd, Hyperasymptotics and the linear boundary layer problem: Why asymptotic series diverge. SIAM Rev. **47**, 553–575 (2005)

[58] B. Braaksma, Phantom ducks and models of excitability. J. Dyn. Differ. Equ. **4**, 485–513 (1992)

[59] F. Brauer, J. Nohel, *The Qualitative Theory of Ordinary Differential Equations: An Introduction* (Benjamin, New York, 1969)

[60] N.I. Brish, On boundary value problems for the equation $\varepsilon y'' = f(x, y, y')$ for small ε. Dokl. Akad. Nauk SSSR **95**, 429–432 (1954)

[61] A. Busemann, Ludwig Prandtl, 1875–1953. Biographical Memoirs Fellows R. Soc. **5**, 193–205 (1960)

[62] A.W. Bush, *Perturbation Methods for Engineers and Scientists* (CRC Press, Boca Raton, FL, 1992)

[63] V.F. Butuzov, A.B. Vasil'eva, N.N. Nefedov, Asymptotic theory of contrast structures (review). Autom. Remote Control **58**, 1068–1091 (1997)

[64] V.F. Butuzov, N.N. Nefedov, K.R. Schneider, Singularly perturbed boundary value problems for systems of Tikhonov's type in the case of exchange of stability. J. Differ. Equ. **159**, 427–446 (1999)

[65] V.F. Butuzov, N.N. Nefedov, L. Recke, K.R. Schneider, On a singularly perturbed initial value problem in case of a double root of the degenerate equation. Nonlinear Anal. **83**, 1–11 (2012)

[66] P.F. Byrd, M.D. Friedman, *Handbook of Elliptic Integrals for Engineers and Scientists* (Springer, Berlin, 1954)

[67] M. Canalis-Durand, Formal expansion of van der Pol equation canard solutions are Gevrey, in *Dynamic Bifurcations*, ed. by E. Benoît, vol. 1493 of *Lecture Notes in Math.* (Springer, Berlin, 1991), pp. 29–39

[68] M. Canalis-Durand, J.P. Ramis, R. Schaefke, Y. Sibuya, Gevrey solutions of singularly perturbed differential equations. J. Reine Angew Math. **518**, 95–129 (2000)

[69] D.M. Cannell, *George Green: Mathematician and Physicist, 1793–1843: The Background to his Life and Work* (Athlone Press, London, 1993)

[70] G. Carrier, Perturbation methods, in *Handbook of Applied Mathematics*, 2nd edn., ed. by C.E. Pearson (Van Nostrand Reinhold, New York, 1990), pp. 747–814

[71] G. Carrier, C. Pearson, *Ordinary Differential Equations* (Blaisdell, Waltham, MA, 1968)

[72] G.F. Carrier, Singular perturbation theory and geophysics. SIAM Rev. **12**, 175–193 (1970)

[73] A.-L. Cauchy, Sur l'emploi légitime des séries divergentes. Comp. Rend. **17**, 370–376 (1843)

[74] L. Cesari, *Asymptotic Behavior and Stability Problems in Ordinary Differential Equations*, 2nd edn. (Springer, Berlin, 1963)

[75] I. Chang, *Thread of the Silkworm* (Basic Books, New York, 1995)

[76] K.W. Chang, F.A. Howes, *Nonlinear Singular Perturbation Phenomena: Theory and Application* (Springer, New York, 1984)

[77] P.B. Chapman, *A Short Course in a Method for Solving Dynamical Systems and Other Related Problems* (Unpublished)

[78] E. Charpentier, E. Ghys, A. Lesne (eds.), *The Scientific Legacy of Poincaré* (Amer. Math. Soc., Providence, RI, 2010)

[79] J. Chen, R.E. O'Malley, Jr., Multiple solutions of a singularly perturbed boundary value problem arising in chemical reactor theory. Lecture Notes in Math. **280**, 314–319 (1972)

[80] L.-Y. Chen, N. Goldenfeld, Y. Oono, Renormalization group theory for global asymptotic analysis. Phys. Rev. Lett. **73**, 1311–1315 (1994)

[81] L.-Y. Chen, N. Goldenfeld, Y. Oono, Renormalization group and singular perturbations: Multiple-scales, boundary layers, and reductive perturbation theory. Phys. Rev. E **54**(1), 376–394 (1996)

[82] H. Cheng, The renormalized two-scale method. Stud. Appl. Math **113**, 381–387 (2004)

[83] H. Cheng, *Advanced Analytic Methods in Applied Mathematics, Science, and Engineering* (LuBan Press, Boston, 2007)

[84] H. Cheng, T.-T. Wu, An aging spring. Stud. Appl. Math. **49**, 183–185 (1970)

[85] H. Chiba, C^1 approximation of vector fields based on the renormalization group method. SIAM J. Appl. Dyn. Syst. **7**(3), 895–932 (2008)

[86] H. Chiba, Simplified renormalizaton group method for ordinary differential equations. *J. Differ. Equ.* **246**, 1991–2019 (2009)

[87] H. Chiba, Extension and unification of singular perturbation methods for ODEs based on the renormalization group method. SIAM J. Dyn. Syst. **8**, 1066–1115 (2009)

[88] C. Chicone, Inertial and slow manifolds for delay equations with small delays. J. Differ. Equ. **190**, 364–406 (2003)

[89] J.A. Cochran, Problems in Singular Perturbation Theory, PhD thesis, Stanford University, Stanford, CA, 1962

[90] E.A. Coddington, N. Levinson, A boundary value problem for a nonlinear differential equation with a small parameter. Proc. Am. Math. Soc. **3**, 73–81 (1952)

[91] E.A. Coddington, N. Levinson, *Theory of Ordinary Differential Equations* (McGraw-Hill, New York, 1955)

[92] J.D. Cole, *Perturbation Methods in Applied Mathematics* (Blaisdell, Waltham, MA, 1968)

[93] J.D. Cole, The development of perturbation theory at GALCIT. SIAM Rev. **36**, 425–430 (1994)

[94] D. Colton, Arthur Erdélyi. Bull. Lond. Math. Soc. **11**, 191–207 (1979)

[95] C. Comstock, The Poincaré-Lighthill perturbation technique and its generalizations. SIAM Rev. **14**, 433–446 (1972)

[96] W.A. Coppel, *Stability and Asymptotic Behavior of Differential Equations* (Heath, Boston, 1965)

[97] V.T. Coppola, R.H. Rand, Averaging using elliptic functions: approximations of limit cycles. Acta Mech. **81**, 125–142 (1990)

[98] E.T. Copson, *Asymptotic Expansions* (Cambridge University Press, Cambridge, 1965)

[99] J. Cornwell, *Hitler's Scientists* (Viking, New York, 2003)

[100] J.G. van der Corput, Introduction to the neutrix calculus. J. d'Analyse Math. **7**, 281–398 (1959)

[101] O. Costin, *Asymptotics and Borel Summability* (CRC Press, Boca Raton, FL, 2009)

[102] R. Courant, D. Hilbert, *Methoden der mathematischen Physik I, II* (Verlag J. Springer, Berlin, 1924, 1937)

[103] R. Courant, D. Hilbert, *Methods of Mathematical Physics*, vol. 1 (Interscience, New York, 1953)

[104] J. Cousteix, J. Mauss, *Asymptotic Analysis and Boundary Layers* (Springer, Berlin, 2007)

[105] J. Cronin, *Mathematical Aspects of Hodgkin-Huxley Neuron Theory* (Cambridge University Press, Cambridge, 1987)

[106] J. Cronin, Analysis of cellular oscillations, in *Analyzing Multiscale Phenomena Using Singular Perturbation Methods*, ed. by J. Cronin and R.E. O'Malley, Jr. (Amer. Math. Soc., Providence, RI, 1999), pp. 133–150

[107] C.F. Curtiss, J.O. Hirschfelder, Integration of stiff equations. Proc. Natl. Acad. Sci. **38**, 235–243 (1952)

[108] G. Dahlquist, On transformations of graded matrices with applications to stiff ODEs. Numer. Math. **47**, 363–385 (1985)

[109] G. Dahlquist, L. Edsberg, G. Skollermo, G. Soderlind, Are the numerical methods and software satisfactory for chemical kinetics? *Numerical Integration of Differential Equations and Large Linear Systems*, vol. 968 of *Lecture Notes in Math.* (Springer, New York, 1982), pp. 149–164

[110] O. Darrigol, *Worlds of Flow: A History of Hydrodynamics from the Bernoullis to Prandtl* (Oxford University Press, Oxford, 2005)

[111] J.W. Dauben, *Abraham Robinson: The Creation of Nonstandard Analysis, a Personal and Mathematical Odyssey* (Princeton University Press, Princeton, NJ, 1995)

[112] C. De Coster, P. Habets, *Two-point Boundary Value Problems: Lower and Upper Solutions* (Elsevier, Amsterdam, 2006)

[113] P. De Maesschalck, Ackerberg-O'Malley resonance in boundary value problems with a turning point of any order. Commun. Pure Appl. Anal. **6**, 311–333 (2007)

[114] P. De Maesschalck, On maximum bifurcation delay in real planar singularly perturbed vector fields. Nonlinear Anal. **68**, 547–576 (2008)

[115] L. Debnath, *Sir James Lighthill and Modern Fluid Dynamics* (Imperial College Press, London, 2008)

[116] M. Desroches, J. Guckenheimer, B. Krauskopf, C. Kuehn, H. Osinga, M. Wechselberger, Mixed-mode oscillations with multiple time scales. SIAM Rev. **54**, 211–288 (2012)

[117] R.L. Devaney, Singular perturbations of complex polynomials. Bull. Am. Math. Soc. **50**, 391–429 (2013)

[118] R.E.L. De Ville, A. Harkin. M. Holzer, K. Josić, T.J. Kaper, Analysis of a renormalization group method and normal form theory for perturbed ordinary differential equations. Phys. D *237*(8), 1029–1052 (2008)

[119] L. Dieci, M.R. Osborne, R.D. Russell, A Riccati transformation method for solving linear BVPs. I: Theoretical aspects, II: Computational aspects. SIAM J. Numer. Anal. **25**, 1055–1092 (1988)

[120] F. Diener, M. Diener, *Nonstandard Analysis in Practice* (Springer, Berlin, 1995)

[121] F. Diener, M. Diener, Ducks and rivers: three existence results, in *Nonstandard Analysis in Practice*, ed. by F. Diener, M. Diener (Springer, Berlin, 1995), pp. 205–224

[122] M. Diener, The canard unchained or how fast/slow dynamical systems bifurcate. Math. Intelligencer **6**(3), 38–49 (1984)

[123] R.B. Dingle, *Asymptotic Expansions: Their Derivation and Interpretation* (Academic Press, London, 1973)

[124] F.W. Dorr, S.V. Parter, L.F. Shampine, Applications of the maximum principle to singular perturbation problems. SIAM Rev. **15**, 43–88 (1973)

[125] P.G. Drazin, N. Riley, *The Navier-Stokes Equations: A Classification of Flows and Exact Solutions*, vol. 334 of *London Math. Soc. Lecture Note Series* (Cambridge University Press, Cambridge, 2006)

[126] H.L. Dryden, Fifty years of boundary layer theory and experiment. Science **121**, 375–380 (1955)

[127] F. Dumortier, R. Roussarie, Canard cycles and center manifolds. Memoirs Am. Math. Soc. **577** (1996)

[128] N.S. Dvortsina, *Ilia Mikhailovich Lifshitz, 1917–1982*. Nauka, Moscow (1989). In Russian

[129] G. Dyson, *Turing's Cathedral: The Origins of the Digital Universe* (Pantheon Books, New York, 2012)

[130] Weinan E., *Principles of Multiscale Modeling* (Cambridge University Press, Cambridge, 2011)

[131] M. Eckert, *The Dawn of Fluid Dynamics* (Wiley-VCH, Weinheim, 2006)

[132] W. Eckhaus, *Matched Asymptotic Expansions and Singular Perturbations* (North-Holland, Amsterdam, 1973)

[133] W. Eckhaus, *Asymptotic Analysis of Singular Perturbations* (North-Holland, Amsterdam, 1979)

[134] W. Eckhaus, Relaxation oscillations including a standard chase on French ducks, in *Asymptotic Analysis II*, vol. 985 of *Lecture Notes in Math*, ed. by F. Verhulst (Springer, Berlin, 1983), pp. 449–494

[135] W. Eckhaus, On modulation equations of Ginzburg-Landau type, in *ICIAM 91*, ed. by R.E. O'Malley, Jr. (SIAM, Philadelphia, 1992), pp. 83–98

[136] W. Eckhaus, Fundamental concepts of matching. SIAM Rev. **36**, 431–439 (1994)

[137] W. Eckhaus, *Witus en de jaren van angst: Ein reconstructie*. (Bas Lubberhuizen, Amsterdam, 1997). In Dutch

[138] W. Eckhaus, E.M. de Jager, Asymptotic solutions of singular perturbation problems for linear differential equations of elliptic type. Arch. Ration. Mech. Anal. **23**, 26–86 (1966)

[139] S.-I. Ei, K. Fujii, T. Kunihiro, Renormalization-group method for reduction of evolution equations; invariant manifolds and envelopes. Ann. Phys. **280**, 236–298 (2000)

[140] M. Epple, A. Karachalios, V. Remmert, Aeronautics and mathematics in National Socialist Germany and Fascist Italy: A comparison of research institutes. OSIRIS **20**, 131–158 (2005)

[141] A. Erdélyi, *Asymptotic Expansions* (Dover, New York, 1956)

[142] G.B. Ermentrout, D.H. Terman, *Mathematical Foundations of Neuroscience* (Springer, New York, 2010)

[143] T. Erneux, *Applied Delay Differential Equations* (Springer, New York, 2009)

[144] R. Estrada, R.P. Kanwal, *A Distributional Approach to Asymptotics: Theory and Applications*, 2nd edn. (Birkhäuser, Boston, 2002)

[145] M.V. Fedoryuk, *Asymptotic Analysis* (Springer, New York, 1994)

[146] N. Fenichel, Geometric singular perturbation theory for ordinary differential equations. J. Differ. Equ. **15**, 77–105 (1979)

[147] N. Fenichel, Global uniqueness in geometric singular perturbation theory. SIAM J. Math. Anal. **16**, 1–6 (1985)

[148] J.E. Flaherty, R.E. O'Malley, Jr., Numerical methods for stiff systems of two-point boundary value problems. SIAM J. Sci. Stat. Comput. **5**, 865–886 (1984)

[149] L. Flatto, N. Levinson, Periodic solutions of singularly perturbed systems. J. Ration. Mech. Anal. **4**, 943–950 (1955)

[150] I. Flügge-Lotz, W. Flügge, Ludwig Prandtl in the nineteen-thirties: reminiscenses. Ann. Rev. Fluid Mech. **5**, 1–9 (1973)

[151] W.B. Ford, *Studies in Divergent Series and Summability* (Macmillan, New York, 1916)

[152] N.D. Fowkes, A singular perturbation method. Part I. Quart. Appl. Math. **XXVI**, 57–69 (1968)

[153] N.D. Fowkes, J.P.O. Silberstein, John Mahony, 1929–1992. Hist. Record Aust. Sci. **10**, 265–291 (1995)

[154] A.C. Fowler, G. Kember, S.G.B. O'Brien, Small exponent asymptotics. IMA J. Appl. Math. **64**, 23–38 (2000)

[155] L.E. Fraenkel, On the method of matched asymptotic expansions. Proc. Camb. Phil. Soc. **65**, 209–284 (1969)

[156] L.S. Frank, On a singularly perturbed eigenvalue problem in the theory of elastic rods. SIAM J. Math. Anal. **21**, 1245–1263 (1990)

[157] L.S. Frank, *Singular Perturbations in Elasticity Theory* (IOS Press, Amsterdam, 1997)

[158] S. Franz, H.-G. Roos, The capriciousness of numerical methods for singular perturbations. SIAM Rev. **53**, 157–173 (2011)

[159] K.-O. Friedrichs, The edge effect in bending and buckling with large deflections, in *Nonlinear Problems in Mechanics of Continua*, vol. 1 of *Proc. Symp. Appl. Math.*, ed. by E. Reissner, W. Prager, J.J. Stoker (Amer. Math. Soc., New York, 1949), pp. 188–193

[160] K.-O. Friedrichs, *Special Topics in Fluid Mechanics* (New York University, New York, 1953). Lecture Notes

[161] K.-O. Friedrichs, Asymptotic phenomena in mathematical physics. Bull. Am. Math. Soc. **61**, 367–381 (1955)

[162] K.-O. Friedrichs, J.J. Stoker, The nonlinear boundary value problem of the buckled plate. Am. J. Math. **63**, 839–888 (1941)

[163] K.-O. Friedrichs, W. Wasow, Singular perturbations of nonlinear oscillations. Duke Math. J. **13**, 367–381 (1946)

[164] N. Fröman, P.O. Fröman, *Physical Problems Solved by the Phase-Integral Method* (Cambridge University Press, Cambridge, 2002)

[165] A. Fruchard, R. Schäfke, *Composite asymptotic expansions and turning points of singularly perturbed ordinary differential equations*, vol. 2066 of *Lect. Notes in Math.* (Springer, Berlin, 2012)

[166] C. Gavin, A. Pokrovskii, M. Prentice, V. Sobolev, Dynamics of a Lotka-Volterra type model with applications to marine phage population dynamics. J. Phys. Conf. Ser. **55**, 80–95 (2006)

[167] C.W. Gear, T.J. Kaper, I.G. Kevrekidis, A. Zagaris, Projecting to a slow manifold: singularly perturbed systems and legacy codes. SIAM J. Appl. Dyn. Syst. **4**, 711–732 (2005)

[168] P. Germain, The 'new' mechanics of fluids of Ludwig Prandtl, in *Ludwig Prandtl, ein Führer in der Strömungslehre*, ed. by G.E.A. Meier (Vieweg, Braunschweig, 2000), pp. 31–40

[169] P. Germain, My discovery of mechanics. In G.A. Maugin, R. Drouot, and F. Sidoroff (eds.), *Continuum Thermomechanics: the Art and Science of Modelling Material Behaviour* (Kluwer, Dordrecht, 2000), pp. 1–24

[170] H. Glauert, *The Elements of Aerofoil and Airscrew Theory* (Cambridge University Press, Cambridge, 1926)

[171] R. Gobbi and R. Spigler, Comparing Shannon to autocorrelation-based wavelets for solving singularly perturbed boundary value problems. BIT **52**(1), 21–43 (2012)

[172] N. Goldenfeld, The renormalization group far from equilibrium: Singular perturbations, pattern formation, and hydrodynamics. Talk, University of Washington, 2010

[173] A.L. Gol'denveizer, *Theory of Elastic Thin Shells* (Pergamon Press, New York, 1961)

[174] D.J. Goldhagen, *Hitler's Willing Executioners: Ordinary Germans and the Holocaust* (Knopf, New York, 1996)

[175] S. Goldstein (ed.), *Modern Developments in Fluid Dynamics* (Oxford University Press, Oxford, 1938) Two volumes

[176] S. Goldstein, Fluid mechanics in the first half of this century. Ann. Rev. Fluid Mech. **1**, 1–28 (1969)

[177] M.H. Gorn, *The Universal Man: Theodore von Kármán's Life in Aeronautics* (Smithsonian Inst. Press, Washington, DC, 1992)

[178] D.A. Goussis, The role of slow system dynamics in predicting the degeneracy of slow invariant manifolds: The case of vdP relaxation oscillations. Phys. D **248**, 16–32 (2013)

[179] J. Grasman, *Asymptotic Methods for Relaxation Oscillations and Applications* (Springer, New York, 1987)

[180] J. Grasman, O.A. van Herwaarden, *Asymptotic Methods for the Fokker-Planck Equations and the Exit Problem in Applications* (Springer, Berlin, 1999)

[181] J. Grasman, B.J. Matkowsky, A variational approach to singularly perturbed boundary value problems for ordinary and partial differential equations with turning points. SIAM J. Appl. Math. **32**, 588–597 (1977)

[182] J. Gray, *Henri Poincaré: A Scientific Biography* (Princeton University Press, Princeton, NJ, 2012)

[183] G. Green, *An Essay on the Applications of Mathematical Analysis to the Theories of Electricity and Magnetism*. Printed for the author by T. Wheelhouse, Nottingham, 1828

[184] W.M. Greenlee, R.E. Snow, Two-timing on the half line for damped oscillator equations. J. Math. Anal. Appl. **51**, 394–428 (1975)

[185] P.P.N. de Groen, The nature of resonance in a singular perturbation problem of turning point type. SIAM J. Math. Anal. **11**, 1–22 (1980)

[186] P.P.N. de Groen, The singularly perturbed turning point problem: A spectral analysis, in *Singular Perturbations and Asymptotics*, ed. by R.E. Meyer, S.V. Parter (Academic Press, New York, 1980), pp. 149–172

[187] Z.-M. Gu, N.N. Nefedov, R.E. O'Malley, Jr., On singular singularly-perturbed initial value problems. SIAM J. Appl. Math. **49**, 1–25 (1989)

[188] J. Guckenheimer, P. Holmes, *Nonlinear Oscillations, Dynamical Systems, and Bifurcations of Vector Fields* (Springer, New York, 1983)

[189] S. Haber, N. Levinson, A boundary value problem for a singularly perturbed differential equation. Proc. Am. Math. Soc. **6**, 866–872 (1955)

[190] R. Haberman, *Applied Partial Differential Equations with Fourier Series and Boundary Value Problems*, 5th edn. (Pearson Prentice-Hall, Upper Saddle River, NJ, 2012)

[191] J. Hadamard, *An Essay on the Psychology of Invention in the Mathematical Field* (Princeton University Press, Princeton, NJ, 1945)

[192] E. Hairer, G. Wanner, *Solving Ordinary Differential Equations II: Stiff and Differential-Algebraic Problems*, 2nd revised edn. (Springer, Berlin, 2010)

[193] G.H. Handelman, J.B. Keller, R.E. O'Malley, Jr., Loss of boundary conditions in the asymptotic solution of linear ordinary differential equations, I: Eigenvalue problems. Comm. Pure Appl. Math. **21**, 243–266 (1968)

[194] G.H. Hardy, *Orders of Infinity: the "Infinitärcalcül" of Paul DuBois-Reymond* (Cambridge University Press, Cambridge, 1910)

[195] G.H. Hardy, *A Mathematician's Apology* (Cambridge University Press, Cambridge, 1940)

[196] G.H. Hardy, *Divergent Series* (Oxford University Press, Oxford, 1949)

[197] W.A. Harris, Jr., Singular perturbations of two-point boundary value problems for systems of ordinary differential equations. Arch. Ration. Mech. Anal. **5**, 212–225 (1960)

[198] S.P. Hastings, J.B. McLeod, *Classical Methods in Ordinary Differential Equations: with Applications to Boundary Value Problems* (Amer. Math. Soc, Providence, 2012)

[199] W.D. Hayes, R.F. Probstein, *Hypersonic Flow Theory* (Academic Press, New York, 1959)

[200] J. Heading, *An Introduction to the Phase-Integral Methods* (Methuen, London, 1962)

[201] G. Hek, Geometric singular perturbation theory in biological practice. J. Math. Biol. **60**, 347–386 (2010)

[202] P.W. Hemker, *A Numerical Study of Stiff Two-Point Boundary Value Problems* (Mathematical Centre, Amsterdam, 1977)

[203] R. Hersh, Under-represented then over-represented: A memoir of Jews in American mathematics. Coll. Math. J. **41**(1), 2–9 (2010)

[204] R. Hersh, V. John-Steiner, *Loving + Hating Mathematics* (Princeton University Press, Princeton, NJ, 2011)

[205] E. Hille, *Analytic Function Theory*, vol. 2 (Ginn, Boston, 1962)

[206] E.J. Hinch, *Perturbation Methods* (Cambridge University Press, Cambridge, 1991)

[207] E.H. Hirschel, H. Prem, G. Madelang (eds.), *Aeronautical Research in Germany: From Lilienthal until Today* (Springer, Berlin, 2004)

[208] M.H. Holmes, *Introduction to the Foundations of Applied Mathematics* (Springer, New York, 2009)

[209] M.H. Holmes, *Introduction to Perturbation Methods*, 2nd edn. (Springer, New York, 2013)

[210] G. Holton, *Thematic Origins of Scientific Thought: Kepler to Einstein* (Harvard University Press, Cambridge, 1973)

[211] M. Holzer, T.J. Kaper, An analysis of the renormalization group method for asymptotic expansions with logarithmic switchback terms. Adv. Differ. Equ. **19**(3/4), 245–282 (2014)

[212] F.C. Hoppensteadt, Singular perturbations on the infinite interval. Trans. Am. Math. Soc. **123**, 521–535 (1966)

[213] F.C. Hoppensteadt, *Analysis and Simulation of Chaotic Systems*, 2nd edn. (Springer, New York, 2000)

[214] F.C. Hoppensteadt, *Quasi-static State Analysis of Differential, Difference, Integral, and Gradient Systems*, vol. 21 of *Courant Lecture Notes* (Amer. Math. Soc, Providence, RI, 2010)

[215] R.A. Horn, C.R. Johnson, *Topics in Matrix Analysis* (Cambridge University Press, Cambridge, 1991)

[216] J. Horvath (ed.), *A Panorama of Hungarian Mathematics in the Twentieth Century* (Springer, Berlin, 2006)

[217] F.A. Howes, Boundary-interior layer interactions in nonlinear singular perturbation theory. Memoirs Am. Math. Soc. **203**, 1–108 (1978)

[218] F.A. Howes, Some singularly perturbed superquadratic boundary value problems whose solutions exhibit boundary and shock layer behavior. Nonlinear Anal. Theory Methods Appl. **4**, 683–698 (1980)

[219] C.J. Howls, Exponential asymptotics and boundary value problems: keeping both sides happy at all orders. Proc. R. Soc. Lond. A **466**, 2771–2794 (2010)

[220] P.-F. Hsieh, Y. Sibuya, *Basic Theory of Ordinary Differential Equations* (Springer, New York, 1999)

[221] A.M. Il'in, *Matching of Asymptotic Expansions of Solutions of Boundary Value Problems* (Amer. Math. Soc, Providence, RI, 1992)

[222] G. Israel, Technical innovation and new mathematics: van der Pol and the birth of nonlinear dynamics, in *Technological Concepts and Mathematical Models in the Evolution of Modern Engineering Systems*, ed. by M. Lucertini, A.M. Gasca, F. Nicolò (Birkhäuser Verlag, Basel, 2004), pp. 52–78

[223] A. Jacobsen, *Operation Paperclip: The Secret Intelligence Program that Brought Nazi Scientists to America* (Little Brown and Company, New York, 2014)

[224] E.M. de Jager, Jiang Furu, *The Theory of Singular Perturbations* (Elsevier, Amsterdam, 1996)

[225] I. James, *Driven to Innovate: A Century of Jewish Mathematicians and Physicists* (Peter Lang, Oxford, 2009)

[226] R.S. Johnson, *Singular Perturbation Theory: Mathematical and Analytical Techniques with Applications to Engineering* (Springer, New York, 2005)

[227] C.K.R.T. Jones, A.I. Khibnik, editors., *Multiple-Time-Scale Dynamical Systems* (Springer, New York, 2001)

[228] D.S. Jones, Arthur Erdélyi 1908–1977. Biographical Memoirs Fellows R. Soc. **25**, 267–286 (1979)

[229] D.S. Jones, *Introduction to Asymptotics: A Treatment using Nonstandard Analysis* (World Scientific, Singapore, 1997)

[230] D.W. Jordan, P. Smith, *Nonlinear Ordinary Differential Equations*, 4th edn. (Oxford University Press, Oxford, 2007)

[231] P.B. Kahn, Y. Zarmi, *Nonlinear Dynamics: Exploration through Normal Forms* (Wiley, New York, 1998)

[232] E. Kamke, *Differentialgleichungen Lösungsmethoden und Lösungen* (Becker and Erler, Leipzig, 1944)

[233] T.J. Kaper, An introduction to geometric methods and dynamical systems theory for singular perturbation problems, in *Analyzing Multiscale Phenomena using Singular Perturbation Methods*, ed. by J. Cronin, R.E. O'Malley, Jr. (Amer. Math. Soc., Providence, 1999), pp. 85–131

[234] A.K. Kapila, *Asymptotic Treatment of Chemically Reacting Systems* (Pitman, London, 1983)

[235] S. Kaplun, Low Reynolds number flow past a circular cylinder. J. Math. Mech. **6**, 595–603 (1957)

[236] M.J. Kaplun, *My Son Saul: Saul Kaplun, July 3,1924–February 13, 1964 in Memoriam* (Shengold, New York, 1965)

[237] S. Kaplun, *Fluid Mechanics and Singular Perturbations*. P. A. Lagerstrom, L. N. Howard, and C. S. Liu, editors. (Academic Press, New York, 1967)

[238] T. von Kármán, The engineer grapples with nonlinear problems. Bull. Am. Math. Soc. **46**, 615–683 (1940)

[239] T. von Kármán, *Aerodynamics: Selected Topics in Light of their Historical Development* (Cornell University Press, Ithaca, NY, 1954)

[240] T. von Kármán, M.A. Biot, *Mathematical Methods in Engineering* (McGraw-Hill, New York, 1940)

[241] T. von Kármán, L. Edson, *The Wind and Beyond: Theodore von Kármán, Pioneer in Aviation and Pathfinder in Space* (Little-Brown, Boston, 1967)

[242] W.L. Kath, Slowly-varying phase planes and boundary-layer theory. Stud. Appl. Math. **72**, 221–239 (1985)

[243] T. Kato, *Perturbation Theory for Linear Operators* (Springer, Berlin, 1976)

[244] J.B. Keller, Accuracy and validity of the Born and Rytov approximations. J. Opt. Soc. Am. **59**(8), 1003–1004 (1969)

[245] J.B. Keller, R.M. Lewis, *Asymptotic Methods for Partial Differential Equations: The Reduced Wave Equation and Maxwell's Equation*, vol. 1 of *Surveys in Applied Mathematics*, ed. by J.B. Keller, G. Papanicolaou, D.W. McLaughlin (Plenum, New York, 1995), pp. 1–82

[246] G.C. Kember, A.C. Fowler, J.D. Evans, S.G.B. O'Brien, Exponential asymptotics with a small exponent. Quart. Appl. Math. **LVIII**(3), 561–576 (2000)

[247] J. Kevorkian, The two-variable expansion procedure for the approximate solution of certain nonlinear differential equations, in *Space Mathematics, Part III*, vol. 7 of *Lectures in Applied Mathematics*, ed. by J.B. Rosser (Amer. Math. Soc., Providence, RI, 1966), pp. 206–275

[248] J. Kevorkian, J.D. Cole, *Perturbation Methods in Applied Mathematics* (Springer, New York, 1981)

[249] J. Kevorkian, J.D. Cole, *Multiple Scale and Singular Perturbation Methods* (Springer, New York, 1996)

[250] J.K. Kevorkian, The uniformly valid asymptotic approximations to the solutions of certain non-linear ordinary differential equations. PhD thesis, California Institute of Technology, Pasadena, CA, 1961

[251] H.K. Khalil, *Nonlinear Systems* (Prentice-Hall, Upper Saddle River, NJ, 1996)

[252] E. Kirkinis, Secular series and renormalization group for amplitude equations. Phys. Rev. E **78**(1), 032104 (2008)

[253] E. Kirkinis, The renormalization group: A perturbation method for the graduate curriculum. SIAM Rev. **54**, 374–388 (2012)

[254] E. Kirkinis, R.E. O'Malley, Jr., Renormalization group and multiple scales for the Swift-Hohenberg and Kuramoto-Sivashinski equations. (to appear)

[255] M. Kline, *Mathematical Thought from Ancient to Modern Times* (Oxford University Press, New York, 1972)

[256] P.V. Kokotović, H.K. Khalil, J. O'Reilly, *Singular Perturbation Methods in Control* (Academic Press, London, 1986)

[257] N.J. Kopell, Waves, shocks, and target patterns in an oscillating chemical reagent, in *Nonlinear Diffusion*, ed. by W.E. Fitzgibbon III, H.F. Walker (Pitman, London, 1977), pp. 129–154

[258] N.J. Kopell, L.N. Howard, Plane wave solutions to reaction-diffusion equations. Stud. Appl. Math. **52**, 291–328 (1973)

[259] N.V. Kopteva, E. O'Riordan, Shishkin meshes in the numerical solution of singularly perturbed differential equations. Int. J. Numer. Anal. Model. **7**(3), 393–415 (2010)

[260] I. Kosiuk, P. Szmolyan, Scaling in singular perturbation problems: Blowing up a relaxation oscillator. SIAM J. Appl. Dyn. Syst. **10**, 1307–1343 (2011)

[261] I. Kovacic, M.J. Brennan (eds.), *The Duffing Equation: Nonlinear Oscillators and their Behaviour* (Wiley, Chichester, 2011)

[262] S.G. Krantz, H.R. Parks, *The Implicit Function Theorem: History, Theory, and Applications* (Birkhäuser, Boston, 2002)

[263] H.-O. Kreiss, Difference methods for stiff ordinary differential equations. SIAM J. Numer. Anal. **15**, 21–58 (1978)

[264] H.-O. Kreiss, S.V. Parter, Remarks on singular perturbations with turning points. SIAM J. Math. Anal. **5**, 230–251 (1974)

[265] M. Krupa, P. Szmolyan, Extending geometric singular perturbation theory to nonhyperbolic points–fold and canard points in two dimensions. SIAM J. Math. Anal. **33**(2), 286–314 (2001)

[266] M.D. Kruskal, Asymptotology, in *Mathematical Models in the Physical Sciences*, ed. by S. Drobot (Prentice-Hall, Englewood Cliffs, NJ, 1963), pp. 17–48

[267] N.M. Krylov, N.N. Bogoliubov, *Introduction to Nonlinear Mechanics* (Princeton University Press, Princeton, 1947)

[268] C. Kuehn, *Multiple-Time-Scale Dynamics* (to appear)

[269] H.T. Kung, J.F. Traub, All algebraic functions can be computed fast. J. ACM **25**, 245–260 (1978)

[270] T. Kunihiro, A geometric formulation of the renormalization group method for global analysis. Prog. Theor. Phys. **94**(4), 503–514 (1995)

[271] T. Kunihiro, Renormalization-group resummation of a divergent series of the perturbation wave functions of quantum systems. Prog. Theor. Phys. Supplement No. 131, 459–470 (1998)

[272] Y. Kuramoto, *Chemical Oscillations, Waves, and Turbulence* (Springer, Berlin, 1984)

[273] G.E. Kuzmak, Asymptotic solutions of nonlinear second order differential equations with variable coefficients. J. Appl. Math. Mech. **23**, 730–744 (1959)

[274] J.G.L. Laforgue, Odd-order turning point: Resonance and dynamic metastability, in *III Coloquio sobre Ecuaciones Diferenciales y Aplicaciones, Volumen II*, ed. by A. Domingo Rueda, J. Guinez (Universidad del Zulia, Maracaibo, 1997) pp. 17–23

[275] J.G.L. Laforgue, R.E. O'Malley, Jr., Shock layer movement for Burgers' equation. SIAM J. Appl. Math. **55**, 332–347 (1995)

[276] P.A. Lagerstrom, *Matched Asymptotic Expansions* (Springer, New York, 1988)

[277] J.L. Lagrange, *Analytical Mechanics* (Kluwer, Dordrecht, 1997). Translated from the French edition of 1811

[278] H. Lamb, *Hydrodynamics*, 6th edn. (Cambridge University Press, Cambridge, 1932)

[279] R. Lamour, R. März, C. Tischendorf, *Differential-Algebraic Equations: A Projector Based Analysis* (Springer, Berlin, 2013)

[280] C.G. Lange, On spurious solutions of singular perturbation problems. Stud. Appl. Math. **68**, 227–257 (1983)

[281] G.E. Latta, Singular perturbation problems. PhD thesis, California Institute of Technology, Pasadena, CA, 1951

[282] D. Leavitt, *The Indian Clerk: A Novel* (Bloomsbury, New York, 2007)

[283] N.R. Lebovitz, R.J. Schaar, Exchange of stabilities in autonomous systems. Stud. Appl. Math. **54**, 229–260 (1975)

[284] J.-W. Lee, M.J. Ward, On the asymptotic and numerical analysis of exponentially ill-conditioned singularly perturbed boundary value problems. Stud. Appl. Math. **94**, 271–326 (1995)

[285] A. Leonard, N. Peters, Theodore von Kármán, in *A Voyage Through Turbulence*, ed. by A. Davidson, Y. Kenada, K. Moffatt, K.R. Sreenivasan (Cambridge University Press, Cambridge, 2010), pp. 101–126

[286] H.C. Levey, J.J. Mahony, Resonance in almost linear systems. J. Inst. Math. Appl. **4**, 282–294 (1968)

[287] N. Levinson, Perturbations of discontinuous solutions of nonlinear systems of differential equations. Acta Math. **82**, 71–106 (1951)

[288] Z. Levinson, in *Recountings: Conversations with MIT Mathematicians*, ed. by J. Segel (A. K. Peters, Wellesley, MA, 2009), pp. 1–24

[289] J.H. Lienhard, Ludwig Prandtl, in *Dictionary of Scientific Biography*, *vol. 11*, ed. by C.C. Gillespie (Scribner, New York, 1970), pp. 123–125

[290] M.J. Lighthill, A technique for rendering approximate solutions to physical problems uniformly valid. Phil. Mag. **40**, 1179–1201 (1949)

[291] C.C. Lin, L.A. Segel, *Mathematics Applied to Deterministic Problems in the Natural Sciences* (Macmillan, New York, 1974)

[292] P. Lin, A numerical method for quasilinear singular perturbation problems with turning points. Computing **46**, 155–164 (1991)

[293] P. Lin, R.E. O'Malley, Jr., The numerical solution of a challenging class of turning point problems. SIAM J. Sci. Comput. **25**, 927–941 (2003)

[294] T. Linss, *Layer-Adapted Meshes for Reaction-Convection-Diffusion Problems*, vol. 1985 of *Lecture Notes in Math* (Springer, Heidelberg, 2010)

[295] J.-L. Lions, *Perturbations Singulières dans les Problèmes aux Limites et en Contrôle Optimal*, vol. 1973 of *Lecture Notes in Math* (Springer, Berlin, 1973)

[296] J.E. Littlewood, *A Mathematician's Miscellany* (Methuen, London, 1953)

[297] L.L. Lo, The meniscus on a needle–a lesson in matching. J. Fluid Mech. **132**, 65–78 (1983)

[298] S.A. Lomov, *Introduction to the General Theory of Singular Pertur-bations*, vol. 112 of *Translations of Mathematical Monographs* (Amer. Math. Soc., Providence, RI, 1992)

[299] J. Lorenz, Analysis of difference schemes for a stationary shock problem. SIAM J. Numer. Anal. **21**, 1038–1053 (1984)

[300] R. Lutz, M. Goze, *Nonstandard Analysis: A Practical guide with Applications*, vol. 881 of *Lecture Notes in Math* (Springer, Berlin, 1981)

[301] J. Lützen, *Joseph Liouville 1809–1882: Master of Pure and Applied Mathematics* (Springer, New York, 1990)

[302] A.D. MacGillivray, A method for incorporating transcendentally small terms into the method of matched asymptotic expansions. Stud. Appl. Math. **99**, 285–310 (1997)

[303] A.D. MacGillivray, R.J. Braun, G. Tanoglu, Perturbation analysis of a problem of Carrier's. Stud. Appl. Math. **104**, 293–311 (2000)

[304] S. MacLane, *Saunders MacLane: A Mathematical Autobiography* (A. K. Peters, Wellesley, MA, 2005)

[305] J.J. Mahony, An expansion method for singular perturbation problems. J. Aust. Math. Soc. **2**, 440–463 (1961–1962)

[306] J.J. Mahony, J.J. Shepherd, Stiff systems of ordinary differential equations. II. boundary value problems for completely stiff systems. J. Aust. Math. Soc. B **23**, 136–172 (1981–1982)

[307] V.P. Maslov, *The Complex WKB Method for Nonlinear Equations I: Linear Theory* (Birkhäuser, Basel, 1994)

[308] B.J. Matkowsky, On boundary layer problems exhibiting resonance. SIAM Rev. **17**, 82–100 (1975)

[309] R.M.M. Mattheij, S.W. Rienstra, J.H.M. ten Thije Boonkkamp, *Partial Differential Equations: Modeling, Analysis, Computation* (SIAM, Philadelphia, 2005)

[310] J.A.M. McHugh, An historical survey of ordinary differential equations with a large parameter and turning points. Arch. Hist. Exact Sci. **7**, 277–324 (1971)

[311] J.B. McLeod, S. Sadhu, Existence of solutions and asymptotic analysis of a class of singularly perturbed boundary value problems. Adv. Differ. Equ. **18**(9–10), 825–848 (2013)

[312] J. Medawar, D. Pyke, *Hitler's Gift: The True Story of the Scientists Expelled by the Nazi Regime* (Arcade Publishing, New York, 2001)

[313] H. Mehrtens, V.M. Kingsbury, The Gleichschaltung of mathematical societies in Nazi Germany. Math. Intell. **11**, 48–60 (1989)

[314] G.E.A. Meier (ed.), *Ludwig Prandtl, ein Führer in der Strömungslehre* (Vieweg, Brunsweig, 2000)

[315] G.E.A. Meier, Prandtl's boundary layer concept and the work in Göttingen, in *IUTAM Symposium on One Hundred Years of Boundary Layer Research*, ed. by G.E.A. Meier, K.R. Sreenivasan, H.-J. Heineman (Springer, Dordrecht, 2006)

[316] R.E. Meyer, A view of the triple deck. SIAM J. Appl. Math. **43**, 639–663 (1983)

[317] J.J.H. Miller, E. O'Riordan, G.I. Shishkin, *Fitted Numerical Methods for Singular Perturbation Problems: Error Estimates in the Maximum Norm for Linear Problems in One and Two Dimensions*, revised edn. (World Scientific, Singapore, 2012)

[318] P.D. Miller, *Applied Asymptotic Analysis* (Amer. Math. Soc., Providence, RI, 2006)

[319] N. Minorsky, *Introduction to Nonlinear Mechanics* (J. W. Edwards, Ann Arbor, MI, 1947)

[320] W.L. Miranker, *Numerical Methods for Stiff Equations and Singular Perturbation Methods* (D. Reidel, Dordrecht, 1981)

[321] E.F. Mischenko, N.Kh. Rozov, *Differential Equations with Small Parameters and Relaxation Oscillations* (Plenum, New York, 1980)

[322] R. von Mises, K.-O. Friedrichs, *Fluid Dynamics* (Brown University, Providence, RI, 1942)

[323] E.F. Mishchenko, Yu. S. Kolesov, A. Yu. Kolesov, N. Kh. Rozov, *Asymptotic Methods in Singularly Perturbed Systems* (Consultants Bureau, New York, 1994)

[324] C. Moler, Wilkinson's polynomials, MATLAB CENTRAL-Cleve's corner: Cleve Moler on Mathematics and Computing (March 4, 2013)

[325] J.V. Moloney, A.C. Newell, *Nonlinear Optics* (Westview, Boulder, CO, 2004)

[326] C.S. Morawetz, *Kurt-Otto Friedrichs Selecta* (Birkhäuser, Boston, 1986)

[327] J.A. Morrison, Comparison of the modified method of averaging and the two-variable expansion procedure. SIAM Rev. **8**, 66–85 (1966)

[328] M.P. Mortell, R.E. O'Malley, Jr., A. Pokrovskii, V. Sobolev (eds.), *Singular Perturbations and Hysteresis* (SIAM, Philadelphia, 2005)

[329] J. Moser, Singular perturbation of eigenvalue problems for linear differential equations of even order. Comm. Pure Appl. Math. **8**, 251–278 (1955)

[330] B. Mudavanhu, R.E. O'Malley, Jr., A renormalization group method for nonlinear oscillators. Stud. Appl. Math. **107**, 63–79 (2001)

[331] B. Mudavanhu, R.E. O'Malley, Jr., A new renormalization method for the asymptotic solution of weakly nonlinear vector systems. SIAM J. Appl. Math. **63**, 373–397 (2002)

[332] B. Mudavanhu, R.E. O'Malley, Jr., D.B. Williams, Working with multiscale asymptotics: Solving weakly nonlinear oscillator equations on long-time intervals. J. Eng. Math. **53**, 301–336 (2005)

[333] P.N. Müller, K.-D. Reinsch, R. Bulirsch, On the complete solution of $\varepsilon y'' = y^3$. J. Optim. Theory Appl. **80**, 367–372 (1994)

[334] J. Murdock, Perturbation methods, in *Mathematical Tools for Physicists*, 2nd edn., edited by M. Grinfeld (Wiley-VCH, Weinheim, 2014).

[335] J.A. Murdock, *Perturbations: Theory and Methods* (Wiley-Interscience, New York, 1991)

[336] J.A. Murdock, Some fundamental issues in multiple scale theory. Appl. Anal. **53**, 157–173 (1994)

[337] J.A. Murdock, L.-C. Wang, Validity of the multiple scale method for very long intervals. Z. Angew. Math. Phys. **47**, 760–789 (1996)

[338] J.D. Murray, *Asymptotic Analysis* (Springer, New York, 1984)

[339] J.D. Murray, *Mathematical Biology, I. An Introduction*, 3rd edn. (Springer, New York, 2002)

[340] A. Narang-Siddarth, J. Valasek, *Nonlinear Multiple Time Scale Systems in Standard and Nonstandard Forms: Analysis and Control* (SIAM, Philadelphia, 2014)

[341] A.H. Nayfeh, *Perturbation Methods* (Wiley, New York, 1973)

[342] A.H. Nayfeh, *Problems in Perturbation* (Wiley, New York, 1985)

[343] A.H. Nayfeh, Resolving controversies in the application of the method of multiple scales and the generalized method of averaging. Nonlinear Dyn. **40**, 61–102 (2005)

[344] A.H. Nayfeh, *The Method of Normal Forms*, 2nd, updated and enlarged edn. (Wiley-VCH, Weinheim, 2011)

[345] A.H. Nayfeh, D.T. Mook, *Nonlinear Oscillations* (Wiley-Interscience, New York, 1979)

[346] N.N. Nefedov, K.R. Schneider, On immediate delayed exchange of stabilities and periodic forced canards. Comput. Math. Math. Phys **48**(1), 43–58 (2008)

[347] K. Nickel, Prandtl's boundary layer theory from the viewpoint of a mathematician. Ann. Rev. Fluid Mech. **5**, 405–428 (1973)

[348] K. Nickelsen, A. Hool, G. Graßhoff, *Theodore von Kármán: Flugzeuge für die Welt und eine Stiftung für Bern* (Birkhäuser, Basel, 2004)

[349] K. Nipp, Breakdown of stability in singularly perturbed autonomous systems: I. orbit equations, II. estimates for the solutions and application. SIAM J. Math. Anal. **17**, 512–532 and 1068–1085 (1986)

[350] K. Nipp, An algorithmic approach for solving singularly perturbed initial value problems. Dyn. Rep. **1**, 173–263 (1988)

[351] K. Nipp, D. Stoffer, *Invariant Manifolds in Discrete and Continuous Systems* (European Math. Soc., Zürich, 2013)

[352] B. Noble, M.A. Hussain, Multiple scaling and a related expansion method with applications, in *Lasers, Molecules, and Methods*, ed. by J.O. Hirschfelder, R.L. Wyatt, R.D. Coalson (Wiley, New York, 1989), pp. 83–136

[353] J.A. Nohel, Commentary, in *Selected Papers of Norman Levinson*, vol. 1, ed. by J.A. Nohel, D.H. Sattinger (Birkhäuser, Boston, 1998), pp. 267–287

[354] H. Ockendon, J.R. Ockendon, *Viscous Flow* (Cambridge University Press, Cambridge, 1995)

[355] J.J. O'Connor, E.F. Robertson, *MacTutor History of Mathematics Archive* (University of St. Andrews, Fife, 2013). URL http://www-history.mcs.st-and.ac.uk/

[356] M.A. O'Donnell, Boundary and corner layer behavior in singularly perturbed semilinear systems of boundary value problems. SIAM J. Math. Anal. **15**, 317–332 (1984)

[357] H. Oertel (ed.), *Prandtl-Essentials of Fluid Mechanics*, 3rd edn. (Springer, New York, 2010)

[358] A.B. Olde Daalhuis, Exponential asymptotics, in *Orthogonal Polynomials and Special Functions: Leuven 2002*, vol. 1817 of *Lecture Notes in Mathematics*, ed. by E. Koelink, W. Van Assche (Springer, Berlin, 2003), pp. 211–244

[359] A.B. Olde Daalhuis, S.J. Chapman, J.R. King, J.R. Ockendon, R.H. Tew, Stokes phenomena and matched asymptotic expansions. SIAM J. Appl. Math. **55**, 1469–1483 (1995)

[360] F.W.J. Olver, *Asymptotics and Special Functions* (Academic Press, New York, 1974)

[361] F.W.J. Olver, D.W. Lozier, R.F. Boisvert, C.W. Clark, *Handbook of Mathematical Functions* (Cambridge University Press, Cambridge, 2010)

[362] R.E. O'Malley, Jr., A boundary value problem for certain nonlinear second order differential equations with a small parameter. Arch. Ration. Mech. Anal. **29**, 66–74 (1968)

[363] R.E. O'Malley, Jr., Topics in singular perturbations. Adv. Math. **2**, 365–470 (1968)

[364] R.E. O'Malley, Jr., Boundary value problems for linear systems of ordinary differential equations involving many small parameters. J. Math. Mech. **18**, 835–855 (1969)

[365] R.E. O'Malley, Jr., On boundary value problems for a singularly perturbed equation with a turning point. SIAM J. Math. Anal. **1**, 479–490 (1970)

[366] R.E. O'Malley, Jr., *Introduction to Singular Perturbations* (Academic Press, New York, 1974)

[367] R.E. O'Malley, Jr., Phase plane solutions to some singular perturbation problems. J. Math. Anal. Appl. **54**, 170–218 (1976)

[368] R.E. O'Malley, Jr., *Singular Perturbation Methods for Ordinary Differential Equations* (Springer, New York, 1991)

[369] R.E. O'Malley, Jr., Obituary: Wolfgang Wasow. SIAM News 2–3 (December 1993)

[370] R.E. O'Malley, Jr., *Thinking About Ordinary Differential Equations* (Cambridge University Press, Cambridge, 1997)

[371] R.E. O'Malley, Jr., Mahony's intriguing stiff equations. J. Aust. Math. Soc. B **40**, 469–474 (1999)

[372] R.E. O'Malley, Jr., Singularly perturbed linear two-point boundary value problems. SIAM Rev. **50**, 459–482 (2008)

[373] R.E. O'Malley, Jr., Singular perturbation theory: A viscous flow out of Göttingen. Ann. Rev. Fluid Mech. **42**, 1–17 (2010)

[374] R.E. O'Malley, Jr., L.V. Kalachev, Regularization of nonlinear differential-algebraic equations. SIAM J. Math. Anal. **25**, 615–629 (1994)

[375] R.E. O'Malley, Jr., E. Kirkinis, Examples illustrating the use of renormalization techniques for singularly perturbed differential equations. Stud. Appl. Math. **122**, 105–122 (2009)

[376] R.E. O'Malley, Jr., E. Kirkinis, A combined renormalization group-multiple scale method for singularly perturbed problems. Stud. Appl. Math. **124**, 383–410 (2010)

[377] R.E. O'Malley, Jr., E. Kirkinis, Two-timing and matched asymptotic expansions for singular perturbation problems. Eur. J. Appl. Math. **22**, 613–629 (2011)

[378] R.E. O'Malley, Jr., E. Kirkinis, Variation of parameters and the renormalization group method. Stud. Appl. Math. **133**, (2014)

[379] R.E. O'Malley, Jr., M.J. Ward, Exponential asymptotics, boundary layer resonance, and dynamic metastability, in *Mathematics is for Solving Problems*, ed. by L.P. Cook, V. Roytburd, M. Tulin (SIAM, Philadelphia, 1996), pp. 189–203

[380] R.E. O'Malley, Jr., D.B. Williams, Deriving amplitude equations for weakly-nonlinear oscillators and their generalizations. J. Comp. Appl. Math. **190**, 3–21 (2006)

[381] Y. Oono, *The Nonlinear World: Conceptual Analysis and Phenomenology* (Springer, Tokyo, 2013)

[382] L.A. Ostrovsky, A.S. Potapov, *Modulated Waves: Theory and Applications* (Johns Hopkins, Baltimore, 1999)

[383] K. Oswatitsch, K. Wieghardt, Ludwig Prandtl and his Kaiser-Wilhelm-Institut. Ann. Rev. Fluid Mech. **19**, 1–23 (1987)

[384] C.-H. Ou, R. Wong, Shooting method for nonlinear singularly perturbed boundary value problems. Stud. Appl. Math. **112**, 161–200 (2004)

[385] C.-H. Ou, R. Wong, Exponential asymptotics and adiabatic invariance of a simple oscillator. C. R. Acad. Sci. Paris Ser. I **343**(7), 457–462 (2006)

[386] R.B. Paris, *Hadamard Expansions and Hyperasymptotic Evaluation: An Extension of the Method of Steepest Descents* (Cambridge University Press, Cambridge, 2011)

[387] W. Paulsen, *Asymptotic Analysis and Perturbation Theory* (CRC Press, Boca Raton, FL, 2014)

[388] C.E. Pearson, On a differential equation of boundary layer type. J. Math. and Phys. **47**, 134–154 (1968)

[389] C.E. Pearson, On nonlinear ordinary differential equations of boundary layer type. J. Math. and Phys. **47**, 351–358 (1968)

[390] T.J. Pedley, James Lighthill and his contributions to fluid mechanics. Ann. Rev. Fluid Mech. **33**, 1–41 (2001)

[391] L.M. Perko, Higher-order averaging and related methods for perturbed periodic and quasi-periodic systems. SIAM J. Appl. Math. **17**, 698–724 (1969)

[392] R.B. Platte, L.N. Trefethen, Chebfun: A new kind of numerical computing, in *Progress in Industrial Mathematics at ECMI 2008*, ed. by A.D. Fitt et al. (Springer, Berlin, 2010), pp. 69–82

[393] H. Poincaré, Sur les intégrales irrégulières des équations lineaires. Acta Math. **8**, 295–344 (1886)

[394] H. Poincaré, *Les Méthodes Nouvelles de la Mécanique Céleste I., II* (Gauthier-Villars, Paris, 1892, 1893)

[395] A. Pokrovskii, V. Sobolev, A naive view of time relaxation and hysteresis, in *Singular Perturbations and Hysteresis*, ed. by M.P. Mortell, R.E. O'Malley, Jr., A. Pokrovskii, V. Sobolev (SIAM, Philadelphia, 2005), pp. 1–59

[396] B. van der Pol, On relaxation oscillations. Phil. Mag. **2**, 978–992 (1926)

[397] H. Pollard, *Celestial Mechanics* (Mathematical Association of America, Washington, DC, 1976)

[398] L.S. Pontryagin, Asymptotic behavior of the solutions of systems of differential equations with a small parameter in the highest derivatives. Am. Math. Soc. Transl. (2) **18**, 295–319 (1961)

[399] L. Prandtl, Über Flüssigkeitsbewegung bei sehr kleiner Reibung, in *Verh. III Int. Math. Kongr.* (Teubner, Leipzig, 1905), pp. 484–491 [Translated to English in *Early Developments of Modern Aeronautics*, ed. by J.A.K. Ackroyd, B.P. Axcell, A.I. Rubin, Butterworth-Heinemann, Oxford, 2001]

[400] L. Prandtl, The generation of vortices in fluids of small viscosity. J. R. Aero. Soc. **31**, 720–741 (1927)

[401] L. Prandtl, *Führer durch Strömungslehre* (Vieweg-Verlag, Braunschweig, 1942)

[402] L. Prandtl, Mein Weg zu hydrodynamischen Theorien. Phys. Bl. **4**, 89–92 (1948)

[403] I. Proudman, J.R.A. Pearson, Expansions at small Reynolds numbers for the flow past a sphere and a cylinder. J. Fluid Mech. **2**, 237–262 (1957)

[404] E.D. Rainville, *Infinite Series* (Macmillan, New York, 1967)

[405] J.-P. Ramis, Séries divergentes et théories asymptotiques. In *Panoramas et Synthèses*, vol. 121 (Soc. Math. France, Paris, 1993), pp. 1–74

[406] R.V. Ramnath, *Multiple Scales Theory and Aerospace Applications* (AIAA Education Series, Renton, VA, 2010)

[407] R.V. Ramnath, *Computation and Asymptotics* (Springer, Heidelberg, 2012)

[408] R.H. Rand, Lecture notes on nonlinear vibrations. `http://www.math.cornell.edu/~rand/randdocs`, version 53, 2012

[409] J.W. Baron Rayleigh (F.R.S. Strutt), *The Theory of Sound, Volumes I and II bound as one.* (Dover, New York, 1945)

[410] R. Reeves, *A Force of Nature: The Frontier Genius of Ernest Rutherford* (Atlas Books, New York, 2008)

[411] C. Reid, *Courant in Göttingen and New York: The Story of an Improbable Mathematician* (Springer, New York, 1976)

[412] C. Reid, K.-O. Friedrichs 1901–1982. Math. Intell. **5**(3), 23–30 (1983)

[413] E.L. Reiss, On multivariable asymptotic expansions. SIAM Rev. **13**, 189–196 (1971)

[414] E.L. Reiss, A new asymptotic method for jump phenomena. SIAM J. Appl. Math. **39**, 440–455 (1980)

[415] H. Reissner, Spannungen in Kugelschalen (Kuppeln). *Festschrift Heinrich Müller-Breslau* (A. Kröner, Leipzig, 1912), pp. 181–193

[416] L.G. Reyna, M.J. Ward, On the exponentially slow motion of a viscous shock. Comm. Pure Appl. Math. **48**, 79–120 (1995)

[417] R.G.D. Richardson, Applied mathematics and the present crisis. Am. Math. Month. **50**, 415–423 (1943)

[418] A.J. Roberts, *Modelling Emergent Dynamics in Complex Systems* (to appear)

[419] H.-G. Roos, M. Stynes, L. Tobiska, *Robust Numerical Methods for Singularly Perturbed Differential Equations*, 2nd edn. (Springer, Berlin, 2008)

[420] S. Rosenblat, Asymptotically equivalent singular perturbation problems. Stud. Appl. Math. **55**, 249–280 (1976)

[421] G.C. Rota, *Indiscrete Thoughts* (Birkhäuser, Boston, 1997)

[422] E. Rothe, Über asymptotische Entwicklungen bei Randwertaufgaben der Gleichung $\Delta\Delta u + \lambda^k u = \lambda^k \psi$. Math. Ann. **109**, 267–272 (1933)

[423] E. Rothe, Asymptotic solution of a boundary value problem. Iowa State Coll. J. Sci. **13**, 369–372 (1939)

[424] D.E. Rowe, "Jewish mathematics" at Göttingen in the era of Felix Klein. ISIS **77**, 422–449 (1986)

[425] D.E. Rowe, Felix Klein as Wissenschaftspolitiker, in *Changing Images in Mathematics: From the French Revolution to the New Millennium*, ed. by U. Bottazzini, A. Dahan Dalmedico (Routledge, London, 2001), pp. 69–91

[426] D.E. Rowe, Mathematics in wartime: Private reflections of Clifford Truesdell. Math. Intell. **34**(4), 29–39 (2012)

[427] R. Roy, *Sources in the Development of Mathematics: Series and Products from the Fifteenth to the Twenty-first Century* (Cambridge University Press, Cambridge, 2011)

[428] L.A. Rubenfeld, On a derivative-expansion technique and some comments on multiple scaling in the asymptotic approximation of solutions of certain differential equations. SIAM Rev. **20**, 79–105 (1978)

[429] A.M. Samoilenko, N. N. Bogoliubov and non-linear mechanics. Russ. Math. Surv. **49**(5), 109–154 (1994)

[430] J.A. Sanders, F. Verhulst, J. Murdock, *Averaging Methods in Nonlinear Dynamical Systems*, 2nd edn. (Springer, New York, 2007)

[431] W. Sarlet, On a common derivation of the averaging method and the two-timescale method. Celest. Mech. **17**, 299–311 (1978)

[432] S.S. Sastry, C.A. Desoer, Jump behavior of circuits and systems. IEEE Trans. Circuits and Systems, **28**(12), 1109–1124 (1981)

[433] D. Schattschneider, *M. C. Escher: Visions of Symmetry* (W. H. Freeman, New York, 2004)

[434] D. Schattschneider, The mathematical side of M. C. Escher. Not. Am. Math. Soc. **57**(6), 706–718 (2010)

[435] A. Schissel, The development of asymptotic solutions of linear ordinary differential equations, 1817–1920. Arch. Hist. Exact Sci. **16**, 307–378 (1977)

[436] A. Schissel, The initial development of the WKB solutions of linear second-order ordinary differential equations and their use in the connection problem. Historia Math. **4**, 183–204 (1977)

[437] H. Schlichting, An account of the scientific life of Ludwig Prandtl. Z. für Flugwissenschaften **23**(9), 297–316 (1975)

[438] H. Schlichting, K. Gersten, *Boundary Layer Theory*, 8th edn. (Springer, Berlin, 2000)

[439] C. Schmeiser, Finite deformation of thin beams. Asymptotic and perturbation methods. IMA J. Appl. Math. **34**, 155–164 (1985)

[440] P.J. Schmid, D.S. Henningson, *Stability and Transition in Shear Flows*, vol. 142 of *Applied Mathematical Sciences* (Springer, New York, 2001)

[441] Z. Schuss, *Theory and Application of Stochastic Processes: An Analytical Approach* (Springer, New York, 2010)

[442] L.W. Schwartz, Milton Van Dyke, the man and his work. Ann. Rev. Fluid Mech. **34**, 1–18 (2002)

[443] J.W. Searl, Extensions of a theorem of Erdélyi. Arch. Ration. Mech. Anal. **50**, 127–138 (1973)

[444] W.R. Sears, Some reflections on Theodore von Kármán. J. Soc. Ind. Appl. Math. **13**, 175–183 (1965)

[445] W.R. Sears, Von Kármán: Fluid dynamics and other things. Physics Today **39**(1), 34–39 (1986)

[446] S.L. Segal, *Mathematicians Under the Nazis* (Princeton University Press, Princeton, 2003)

[447] L.A. Segel, L. Edelstein-Keshet, *A Primer on Mathematical Methods in Biology* (SIAM, Philadelphia, 2013)

[448] L.A. Segel, M. Slemrod, The quasi-steady state assumption: A case study in perturbation. SIAM Rev. **31**, 446–477 (1989)

[449] H. Segur, S. Tanveer, H. Levine (eds.), *Asymptotics Beyond All Orders* (Plenum, New York, 1991)

[450] E. Shchepakina, V. Sobolev, M.P. Mortell, *Singular Perturbations: Introduction to system order reduction methods with applications*, vol. 2114 of *Lecture Notes in Math.* (Springer, Berlin, 2014)

[451] J. Shen, M. Han, Canard solution and its asymptotic approximation in a second-order nonlinear singularly-perturbed boundary value problem with a turning point. Comm. Nonlinear Sci. Numer. Simulat. **19**, 2632–2643 (2014)

[452] G.I. Shishkin, L.P. Shishkina, *Difference Methods for Singular Perturbation Problems* (CRC Press, Boca Raton, FL, 2009)

[453] M.A. Shishkova, A discussion of a certain system of differential equations with a small coefficient of the highest-order derivatives. Soviet Math. Dokl. **14**, 483–487 (1973)

[454] C. Shubin, Singularly perturbed integral equations. J. Math. Anal. Appl. **313**, 234–250 (2006)

[455] Y. Sibuya, Uniform simplification in a full neighborhood of a transition point. Memoirs Am. Math. Soc. **149**, vi–106 (1974)

[456] Y. Sibuya, A theorem concerning uniform simplification at a transition point and a problem of resonance. SIAM J. Math. Anal. **12**, 653–668 (1981)

[457] Y. Sibuya, The Gevrey asymptotics in the case of singular perturbations. J. Differ. Equ. **165**, 255–314 (2000)

[458] Y. Sibuya, Formal power series solutions in a parameter. J. Differ. Equ. **190**, 559–578 (2003)

[459] Y. Sibuya, K. Takahasi, On the differential equation $(x+\epsilon u)\frac{du}{dx}+q(x)u-r(x) = 0$. Funkcialaj Ekvacioj **9**, 71–81 (1966)

[460] R. Siegmund-Schultze, *Rockefeller and the Internationalization of Mathematics Between the Two World Wars* (Birkhäuser, Basel, 2001)

[461] R. Siegmund-Schultze, The late arrival of academic applied mathematics in the United States: a paradox, theses, and literature. NTM Z. für Geschichte der Wissenschaften, Technik, und Medizin **11**, 116–127 (2003)

[462] R. Siegmund-Schultze, *Mathematicians Fleeing from Nazi Germany* (Princeton University Press, Princeton, 2009)

[463] L.A. Skinner, *Singular Perturbation Theory* (Springer, New York, 2011)

[464] C.G. Small, *Expansions and Asymptotics for Statistics* (CRC Press, Boca Raton, FL, 2010)

[465] D.R. Smith, The multivariable method in singular perturbation analysis. SIAM Rev. **17**, 221–273 (1975)

[466] D.R. Smith, *Singular-Perturbation Theory: An Introduction with Applications* (Cambridge University Press, Cambridge, 1985)

[467] I.J. Sobey, *Introduction to Interactive Boundary Layer Theory* (Oxford University Press, Oxford, 2000)

[468] V. Sobolev, Canard cascades in biological models. Talk given at Nonlinear Dynamics Workshop in Memory of Alexei Pokrovskii, University College Cork, Ireland, 2011

[469] K. Soetaert, J. Cash, F. Mazzia, *Solving Differential Equations in R* (Springer, Berlin, 2012)

[470] J. Spencer (with L. Florescu), *Asymptopia* (American Mathematical Society, Providence, 2014)

[471] K.R. Sreenivasan, G.I. Taylor: the inspiration beyond the Cambridge school, in *A Voyage through Turbulence*, ed. by P.A. Davidson, Y. Kenada, K. Moffatt, K.R. Sreenivasan (Cambridge University Press, Cambridge, 2011), pp. 127–186

[472] I. Stakgold, *Green's Functions and Boundary Value Problems* (Wiley-Interscience, New York, 1979)

[473] B. Yu. Sternin, V.E. Shatalov, *Borel-Laplace Transform and Asymptotic Theory: Introduction to Resurgent Analysis* (CRC Press, Boca Raton, FL, 1996)

[474] K. Stewartson, d'Alembert's paradox. SIAM Rev. **23**, 308–342 (1981)

[475] T.J. Stieltjes, Recherches sur quelques séries semi-convergentes. Ann. de L'Éc. Norm. Sup. **3**, 201–258 (1886)

[476] J.J. Stoker, *Nonlinear Vibrations in Mechanical and Electrical Systems* (Interscience, New York, 1950)

[477] A. Stubhaug, *Gösta Mittag-Leffler: A Man of Conviction* (Springer, Heidelberg, 2010)

[478] G.G. Szpiro, *Poincaré's Prize: The Hundred-Year Quest to Solve One of Math's Greatest Puzzles* (Penguin, New York, 2008)

[479] I. Tani, History of boundary layer theory. Ann. Rev. Fluid Mech. **9**, 87–111 (1977)

[480] G.I. Taylor, When aeronautical science was young. J. R. Aeronaut. Soc. **70**, 108–113 (1966)

[481] G.I. Taylor, Memories of von Kármán. SIAM Rev. **15**, 447–452 (1973)

[482] G.F.J. Temple, *100 Years of Mathematics: A Personal Viewpoint* (Springer, New York, 1981)

[483] L. Ting, Boundary layer theory to matched asymptotics. Z. Angew. Math. Mech. **80**, 845–855 (2000)

[484] H. Trischler, Self-mobilization or resistance? Aeronautical research and National Socialism, in *Science, Technology and National Socialism*, ed. by M. Renneberg, M. Walker (Cambridge University Press, Cambridge, 1994), pp. 72–87

[485] Y.-W. Tschen, Über das Verhalten der Lösungen einer Folge von Differentialgleichungsproblemen, welche im Limes ansarten. Comp. Math. **2**, 378–401 (1935)

[486] H.-S. Tsien, The Poincaré-Lighthill-Kuo method. Adv. Appl. Mech. **4**, 281–349 (1956)

[487] J. Tucciarone, The development of the theory of summable divergent series from 1880 to 1925. Arch. Hist. Exact Sci. **10**, 1–40 (1973)

[488] H.L. Turrittin, My mathematical expectations, Lecture Notes in Mathematics **312**, 1–22 (1973)

[489] F. Ursell, Integrals with a large parameter. A strong form of Watson's lemma, in Elasticity, Mathematical Methods, and Applications, ed. by G. Eason and R.W. Ogden. (Ellis Horwood, Chichester, 1991)

[490] M. Van Dyke, *Perturbation Methods in Fluid Dynamics* (Academic Press, New York, 1964)

[491] M. Van Dyke, *Perturbation Methods in Fluid Dynamics*, Annotated edn. (Parabolic Press, Stanford, CA, 1975)

[492] M. Van Dyke, Nineteenth-century roots of the boundary-layer idea. SIAM Rev. **36**, 415–424 (1994)

[493] V.S. Varadarajan, *Euler Through Time: A New Look at Old Themes* (Amer. Math. Soc., Providence, RI, 2006)

[494] A.B. Vasil'eva, Asymptotic methods in the theory of ordinary differential equations containing small parameters in front of the higher derivatives. USSR Comp. Math. Math. Phys. **3**, 823–863 (1963)

[495] A.B. Vasil'eva, On the development of singular perturbation theory at Moscow State University and elsewhere. SIAM Rev. **36**, 440–452 (1994)

[496] A.B. Vasil'eva, V.F. Butuzov, L.V. Kalachev, *The Boundary Function Method for Singular Perturbation Problems* (SIAM, Philadelphia, 1995)

[497] A.B. Vasil'eva, V.F. Butuzov, N.N. Nefedov, Singularly perturbed problems with boundary and internal layers. Proc. Steklov Inst. Math. **268**, 258–273 (2010)

[498] A.E.P. Veldman, Matched asymptotic expansions and the numerical treatment of viscous-inviscid interaction. J. Eng. Math **39**, 189–206 (2001)

[499] F. Verhulst, Perturbation theory from Lagrange to van der Pol. Nieuw Archief voor Wiskunde **2**, 428–438 (1984)

[500] F. Verhulst, *Methods and Applications of Singular Perturbations: Boundary Layers and Multiple Timescale Dynamics* (Springer, New York, 2005)

[505] F. Verhulst, Periodic solutions and slow manifolds. Int. J. Bifurcat. Chaos **17**, 2533–2540 (2007)

[502] F. Verhulst, *Henri Poincaré, Impatient Genius* (Springer, New York, 2012)

[503] F. Verhulst, Hunting French ducks in population dynamics. *Proceedings, 12th Conference on Dynamical Systems Theory and Applications, Łódz* (Wydawnictwo Politechniki Łódzkie, Łódz, Poland, 2013)

[504] F. Verhulst, Profits and pitfalls of timescales in asymptotics. SIAM Rev. **56**, (2014)

[505] F. Verhulst, T. Bakri, The dynamics of slow manifolds, J. Indones. Math. Soc. **13**, 73–90 (2007)

[506] M.I. Vishik, L.A. Lyusternik, On the asymptotic behavior of the solutions of boundary value problems for quasilinear differential equations. Dokl. Acad. Nauk SSR **121**, 778–781 (1958)

[507] M.I. Vishik, L.A. Lyusternik, The solution of some perturbation problems for matrices and self-adjoint or non-selfadjoint differential equations I. Russ. Math. Surv. **15**(3), 1–75 (1960)

[508] M.I. Vishik, L.A. Lyusternik, Regular degeneration and boundary layer for linear differential equations with a small parameter. Am. Math. Soc. Transl. (2) **2**(20), 239–364 (1961)

[509] J. Vogel-Prandtl, Ludwig Prandtl, ein Lebensbild: Erinnerungen, Dokumente. Mitteilungen aus dem Max-Planck Inst. für Ström., 107, 1993. An English translation by V. Vasanta Ram is available on the website of the International Centre for Theoretical Physics, Trieste

[510] M.J. Ward, Eliminating indeterminacy in singularly perturbed boundary value problems with translation invariant potentials, Stud. Appl. Math. **87**, 95–134 (1992)

[511] W. Wasow, On the asymptotic solution of boundary value problems for ordinary differential equations containing a parameter. J. Math. and Phys. **23**, 173–183 (1944)

[512] W. Wasow, Asymptotic expansions for ordinary differential equations: Trends and problems, in *Asymptotic Solutions of Differential Equations and their Applications*, ed. by C.H. Wilcox (Academic Press, New York, 1964), pp. 3–26

[513] W. Wasow, *Asymptotic Expansions for Ordinary Differential Equations* (Wiley-Interscience, New York, 1965)

[514] W. Wasow, The capriciousness of singular perturbations. Nieuw Arch. Wisk. **18**, 190–210 (1970)

[515] W. Wasow, *Linear Turning Point Theory* (Springer, New York, 1985)

[516] W. Wasow, *Memories of Seventy Years* (Private printing, 1986)

[517] W.R. Wasow, On boundary layer problems in the theory of ordinary differential equations. PhD thesis, New York University, New York, 1942

[518] E.J. Weniger, Nonlinear sequence transformations for the acceleration and the summation of divergent series. Comp. Phys. Rep. **10**, 189–371 (1989)

[519] R.B. White, *Asymptotic Analysis of Differential Equations* (Imperial College Press, London, 2010)

[520] G.B. Whitham, *Linear and Nonlinear Waves* (Wiley, New York, 1974)

[521] E.T. Whittaker, G.N. Watson, *A Course of Modern Analysis*, 4th edn. (Cambridge University Press, Cambridge, 1952)

[522] S. Wiggins, *Introduction to Applied Nonlinear Dynamical Systems and Chaos*, 2nd edn. (Springer, New York, 2003)

[523] J.H. Wilkinson, The evaluation of zeroes of ill-conditioned polynomials: Part 1. Num. Math. **1**, 150–166 (1959)

[524] D. Willett, On a nonlinear boundary value problem with a small parameater multiplying the highest derivative. Arch. Ration. Mech. Anal. **23**, 276–287 (1966)

[525] R. Wong, *Asymptotic Approximation of Integrals* (Academic Press, New York, 1989)

[526] R. Wong, A panoramic view of asymptotics, in *Foundations of Computational Mathematics, Hong Kong 2008*, ed. by F. Cucker, A. Pinkus, M.J. Todd (Cambridge University Press, Cambridge, 2009), pp. 190–235

[527] R. Wong, Y. Zhao, On the number of solutions to Carrier's problem. Stud. Appl. Math. **120**, 213–245 (2008)

[528] S.L. Woodruff, The use of an invariance condition in the solution of multiple-scale singular perturbation problems: Ordinary differential equations. Stud. Appl. Math. **90**, 225–248 (1993)

[529] S.L. Woodruff, A uniformly-valid asymptotic solution to a matrix system of ordinary differential equations and a proof of validity. Stud. Appl. Math. **94**, 393–413 (1995)

[530] M. Yamaguti, L. Nirenberg, S. Mizohata, Y. Sibuya (eds.), *Mitio Nagumo Collected Papers* (Springer, Tokyo, 1993)

[531] E. Zauderer, *Partial Differential Equations of Applied Mathematics*, 2nd edn. (Wiley, New York, 1989)

[532] A. Zettl, *Sturm-Liouville Theory* (Amer. Math. Soc., Providence, RI, 2005)

[533] R.Kh. Zeytounian, *Navier-Stokes-Fourier Equations: A Rational Asymptotic Modeling Point of View* (Springer, Heidelberg, 2012)

[534] R.Kh. Zeytounian, *Five Decades of Tackling Models for Stiff Fluid Dynamics Problems: A Scientific Autobiography* (Springer, Berlin, 2014)

[535] W. Zhang, H.L. Ho, Y.H. Qian, F.B. Gao, A refined asymptotic perturbation method for nonlinear dynamical systems. Arch. Appl. Mech. **84**, 591–606 (2014)

[536] M. Ziane, On a certain renormalization group method. J. Math. Phys. **41**, 3290–3299 (2000)

[537] G.M. Ziegler, *Do I Count? Stories from Mathematics* (CRC Press, Boca Raton, FL, 2014)

[538] A.K. Zvonkin, M.A. Shubin, Nonstandard analysis and singular perturbations of ordinary differential equations. Russ. Math. Surv. **39**(2), 69–131 (1984)

Index

© Springer International Publishing Switzerland 2014   251
R. O'Malley, *Historical Developments in Singular Perturbations*,
DOI 10.1007/978-3-319-11924-3

Printed in the United States
By Bookmasters